Reproduction in mammals

Book 5: Manipulating reproduction

SECOND EDITION **Reproduction in mammals**

BOOK **5** *Manipulating reproduction*

EDITED BY C. R. AUSTIN

Formerly Fellow of Fitzwilliam College
Emeritus Charles Darwin Professor of Animal Embryology
University of Cambridge

AND R. V. SHORT, FRS

Professor of Reproductive Biology
Monash University, Melbourne, Australia

DRAWINGS BY JOHN R. FULLER

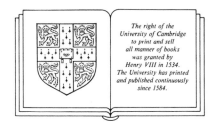

*The right of the
University of Cambridge
to print and sell
all manner of books
was granted by
Henry VIII in 1534.
The University has printed
and published continuously
since 1584.*

Cambridge University Press

Cambridge

London   New York   New Rochelle

Melbourne   Sydney

Published by the Press Syndicate of the University of Cambridge
The Pitt Building, Trumpington Street, Cambridge CB2 1RP
32 East 57th Street, New York, NY 10022, USA
10 Stamford Road, Oakleigh, Melbourne 3166, Australia

First published as *Artificial Control of Reproduction* 1972
Reprinted 1973
Second edition 1986

Printed in Great Britain by the University Press, Cambridge

British Library cataloguing in publication data
Reproduction in mammals.–2nd ed.
Bk. 5, Manipulating reproduction

1. Mammals–Reproduction
I. Austin, C.R.   II. Short, R.V.
599.01′6   QP251

Library of Congress cataloguing in publication data
Reproduction in mammals.

Includes bibliographies and indexes.
Contents: bk. 1. Germ cells and fertilization–
bk. 2. Embryonic and fetal development–[etc.]–
bk. 5. Manipulating reproduction.
1. Mammals–Reproduction–Collected works.   I. Austin,
C. R. (Colin Russell), 1914–   . II. Short, Roger
Valentine, 1930–   . [DNLM: 1. Mammals–Physiology.
2. Reproduction, WQ 205 R4282]
QL739.s.R46 1982   599.01′6   81-18060

ISBN 0 521 307643 hard covers
ISBN 0 521 314968 paperback
(First edition:
ISBN 0 521 085055 hard covers
ISBN 0 521 097134 paperback)

UP

CONTENTS

CONTRIBUTORS TO BOOK 5

C. R. Austin
47 Dixon Road
Buderim, Q 4556, Australia

K. J. Betteridge
Department of Biological Sciences
Ontario Veterinary College
University of Guelph
Guelph, Ontario, N1G 2W1
Canada

J. Cohen
Department of Physiology and Pharmacology
College of Veterinary Medicine
The University of Georgia
Athens, GA 30602, USA

R. G. Edwards, FRS
Bourn Hall Clinic
Bourn
Cambridge CB3 7TR, UK

C. B. Fehilly
Department of Obstetrics and Gynaecology
Righshospitalet
University of Copenhagen
DK–2100 Copenhagen, Denmark

A. McLaren, FRS
MRC Mammalian Development Unit
Wolfson House
4 Stephenson Way
London NW1 2HE, UK

D. M. Potts
Family Health International
Research Triangle Park
North Carolina 27709, USA

R. V. Short, FRS
Department of Physiology
Monash University
Clayton, Victoria 3168, Australia

M. P. Vessey
Department of Community Medicine and General Practice
University of Oxford
Oxford OX1 6HE, UK

PREFACE TO THE SECOND EDITION

In this, our Second Edition of *Reproduction in Mammals*, we are responding to numerous requests for a more up-to-date and rather more detailed treatment of the subject. The First Edition was accorded an excellent reception, but Books 1 to 5 were written some 15 years ago and inevitably there have been many advances on many fronts since then. As before, the manner of presentation is intended to make the subject matter interesting to read and readily comprehensible to undergraduates in the biological sciences, and yet with sufficient depth to provide a valued source of information to graduates engaged in both teaching and research. Our authors have been selected from among the best known in their respective fields.

Book 5 is concerned with the many ways in which we can now manipulate reproductive processes in animals and humans, thanks to our new understanding of hormone action and improved control over early developmental events. We have at our disposal a whole array of synthetic hormone agonists, antagonists and antibodies that can be used at will to stimulate or inhibit fertility in animals and humans alike, so that productivity in livestock can be promoted according to plan and child-bearing becomes more a matter of choice than chance. We can compensate for infertility by *in vitro* fertilization and embryo transfer, and overcome inherent deficiencies by techniques involving embryo manipulation. Existing barriers to the dissemination and application of this new-found knowledge are discussed in some detail, since it is becoming increasingly clear that improvements in the quality of life for many developing countries will be long delayed unless they can meet essential needs and call a halt to runaway population growth.

BOOKS IN THE FIRST EDITION

Books 1–5 are now out of print

Book 6. The evolution of reproduction (1976)
The development of sexual reproduction *S. Ohno*
Evolution of viviparity in mammals *G. B. Sharman*
Selection for reproductive success *P. A. Jewell*
The origin of species *R. V. Short*
Specialization of gametes *C. R. Austin*

Book 7. Mechanisms of hormone action (1979)
Releasing hormones *H. M. Fraser*
Pituitary and placental hormones *J. Dorrington*
Prostaglandins *J. R. G. Challis*
The androgens *W. I. P. Mainwaring*
The oestrogens *E. V. Jensen*
Progesterone *R. B. Heap and A. P. F. Flint*

Book 8. Human sexuality (1980)
The origins of human sexuality *R. V. Short*
Human sexual behaviour *J. Bancroft*
Variant forms of human sexual behaviour *R. Green*
Patterns of sexual behaviour in contemporary society *M. Schofield*
Constraints on sexual behaviour *C. R. Austin*
A perennial morality *G. R. Dunstan*

BOOKS IN THE SECOND EDITION

Book 1. Germ cells and fertilization (1982)
Primordial germ cells and the regulation of meiosis *A. G. Byskov*
Oogenesis and ovulation *T. G. Baker*
The egg *C. R. Austin*
Spermatogenesis and spermatozoa *B. P. Setchell*
Sperm and egg transport *M. J. K. Harper*
Fertilization *J. M. Bedford*

Book 2. Embryonic and fetal development (1982)
The embryo *A. McLaren*
Implantation and placentation *M. B. Renfree*
Sex determination and differentiation *R. V. Short*
The fetus and birth *G. C. Liggins*
Pregnancy losses and birth defects *P. A. Jacobs*
Manipulation of development *R. L. Gardner*

Book 3. Hormonal control of reproduction (1984)
The hypothalamus and anterior pituitary gland *F. J. Karsch*
The posterior pituitary *D. W. Lincoln*
The pineal gland *G. A. Lincoln*
The testis *D. M. de Kretser*
The ovary *D. T. Baird*
Oestrous and menstrual cycles *R. V. Short*
Pregnancy *R. B. Heap and A. P. F. Flint*
Lactation *A. T. Cowie*

Book 4. Reproductive fitness (1984)
Reproductive strategies *R. M. May and D. I. Rubenstein*
Species differences in reproductive mechanisms *R. V. Short*
Genetics and reproduction *R. B. Land*
The environment and reproduction *B. K. Follett*
Reproductive behaviour *E. B. Keverne*
Immunological factors in reproductive fitness *N. J. Alexander and D. J. Anderson*
Reproductive senescence *C. E. Adams*

Book 5. Manipulating reproduction (1986)
Increasing productivity in farm animals *K. J. Betteridge*
Today's and tomorrow's contraceptives *R. V. Short*
Contraceptive needs of the developing world *D. M. Potts*
Benefits and risks of contraception *M. P. Vessey*
Alleviating human infertility *J. Cohen, C. B. Fehilly and R. G. Edwards*
Reproductive options, present and future *A. McLaren*
Barriers to population control *C. R. Austin*

1

Increasing productivity in farm animals

K. J. BETTERIDGE

Horrifying television images have ensured that we are all aware of the misery of the millions of people in the world today who suffer from starvation or malnutrition. It has been estimated that two out of every three people are hungry, most of them in the Third World. We are told, too, that the scale of the problem will continue to rise in parallel with the increase in human population, and we know that the provision of food will always remain the most fundamental of man's challenges.

Television reports also tend to equate famine relief with Western grain. Essential though this is in the short term, we should not forget that stable agricultural systems capable of counteracting poverty and hunger depend on livestock as sources of food and hides, traction and fertilizer, wealth and revenue. The developing world contains over 60 per cent of the world's domestic animals and the potential for improving their productivity is illustrated by figures from the Food and Agriculture Organization of the United Nations (FAO): two-thirds of the world's cattle (including 44 per cent of its dairy cows) and nearly all the buffalo are in developing countries, and yet they produce only 30 per cent of the world's beef and buffalo meat and only 17 per cent of its milk; similarly, little more than one-third of the world's pig meat comes from 57 per cent of the world's pigs to be found in the Third World (Tables 1.1 and 1.2).

Paradoxically, on a world scale, the contribution of animals to man's

Table 1.1. *Distribution of the world's livestock population in 1983*

Livestock	World total (million tonnes)	Distribution (%)	
		Developed countries	Developing countries
Cattle	1225	35	65
Buffalo	124	1	99
Sheep	1137	47	53
Goats	476	6	94
Pigs	774	43	57
Chickens	7063	44	56

(Adapted from *FAO 1983 Production Yearbook*, vol. 37. Food and Agriculture Organization of the UN, Rome, 1984.)

food supply is not entirely, or even mainly, through the provision of meat, milk and eggs. In a closely reasoned analysis of animal production and the world food situation, John Holmes of the University of Melbourne shows that their main contribution to the poor is through the provision of employment, draught power, fertilizer and a cash income with which to buy cereals. Those of us who take mechanized agriculture for granted should remember that about 30 per cent of the entire power requirements for crop production in developing countries comes from animals, and about 66 per cent from men and women themselves. However, the effects of scientific manipulation of animal production are likely to be confined to advanced agricultural systems for a long time yet, and so this chapter will deal primarily with how such manipulation can increase the supply of animal protein for our food.

Why animal protein? Do we humans really need to go through the complexities of dairy farming and the unpleasantness of killing animals in order to survive? Is meat, in fact, just a luxury of the Western diet and an inappropriate answer to world malnutrition? If we do need meat, would it not be better to produce it all from poultry rather than from domesticated mammals? Since it is quite fashionable to presume that it is 'inefficient' to convert plant protein into animal protein before using it to feed ourselves, some answers to these questions should help put the importance of increasing farm animal productivity into perspective.

The importance of animal production

The proportion of the human diet that is made up of food derived from livestock and poultry varies with living standards. In the USA, it represents about 53 per cent of all food consumed and supplies two-thirds of the protein, one-third of the energy, one-half of the fat, and major proportions of minerals, micronutrients and vitamins to the consumer. By contrast, in developing countries, the Centre de Documentation Economique et Sociale Africaine (CEDESA) estimates that '...a bowl of rice, of beans or

Table 1.2. *World livestock production in 1983*

Produce	World total (million tonnes)	Distribution %	
		Developed countries	Developing countries
Beef and buffalo meat	45.6	70	30
Sheep and goat meat	8.2	46	54
Pig meat	54.0	65	35
Cows' milk	454.0	83	17
Hens' eggs	28.7	64	36

(Adapted from *FAO 1983 Production Yearbook*, vol. 37. Food and Agriculture Organization of the UN, Rome, 1984.)

of soya, a cup of milk, a piece of fish, a diet containing a very modest amount of meat, would suffice to turn the mediocre daily fare of many peoples of the world into a highly nutritious diet', and, as we have seen, these modest supplementations of energy and protein are often more likely to be met as an indirect result of animal production. In the USA, the percentage increase in animal production called for between the years 1974 and 2000 to meet forecast levels of consumption is 43 per cent for beef and veal, 36 per cent for pork, 22 per cent for lamb and dairy products, 10 per cent for eggs and 70 per cent for poultry meat. These are formidable, direct goals dictated by consumer preferences in a prosperous society. By contrast, in the Third World, merely adequate diets will depend first on the abolition of poverty, and so the main role of animal production will be to contribute to more stable agricultural economies. The proportion of existing world food supplies that is in the form of livestock is also interesting. Farm animals (including poultry) make up a 40-day reserve for mankind, whereas grain stocks would last for only 27 days. The animals' value to us is related to the type of food that they themselves convert into animal protein.

Livestock that consume forages (mostly ruminants) are of special value because they convert food that is inedible for man into edible animal products. The significance of this is clear when we consider that about two-thirds of the world's agricultural land is in the form of permanent meadows and pastures, about 60 per cent of which are unsuitable for cultivation. In the USA, about 77 per cent of agricultural land is used for grazing by ruminants, defined by Tony Cunha as 'walking protein factories which travel in harvesting their food without the use of fossil fuel'.

The amount of grain fed to food-producing animals instead of directly to man varies with political, economic and sociological conditions. However, even countries that have to import their grain are estimated to feed about 85 per cent of it to animals. It is therefore important to have some idea of how well various classes of livestock make use of it in producing protein for human consumption. Table 1.3 makes this comparison basing feed conversion *only on the parts of the animals' diets that could themselves be used for human food*. The evident advantage of using grain and oilseed meals for feeding ruminants rather than producing pork or poultry meat is again due to the ruminant's ability to make most of its protein from forages and non-protein nitrogen compounds. It is striking that dairy farming actually provides a net gain in protein production when viewed in this way. Tony Cunha also reminds us that comparisons between grain and animal products as human food must take account of the fact that grain is not all consumed as harvested. It requires, for example, 138 kg of wheat to produce 100 kg of wheat flour or 243 kg of oats to produce 100 kg of oatmeal. Each kilogramme of cotton produced is associated with over a kilogramme of cotton-seed oil and meal. Some of the by-products of these processes can only be turned into human food by first feeding them to animals. In many parts of the world, pigs serve a similar purpose by converting waste human food into meat.

It would obviously be invidious to use facts and figures from a feedstuff trade journal to argue that animal production is in any way 'superior' to plant production as a means of feeding the world. However, what they do show is that animals are not as inefficient as is sometimes suggested during polarized arguments between plant scientists and animal scientists. The futility of taking extreme positions in such debates has been discussed in Holmes's analysis to which we referred earlier. He points out that:

> ...both plant and animal production are needed, there is no one 'correct' balance between the two; the balance must vary from region to region depending upon political factors as well as production potential and demand. The development of yet more efficient systems of production and of more precise matching of adapted varieties of plants, animals and agricultural systems with local environments will require an increase in diversity of these plants, animals and systems. An increase in integration of plant and animal production appears to be a more likely outcome than the reverse.

Let there be no doubt, then, that farm animals, and farm mammals in particular, will continue to be vital to our survival. Now let us consider ways and means of increasing the efficiency with which we produce them, and how productivity can be increased as a result of advances in reproductive biology.

Table 1.3. *Natural protein (from grain and oilseed meals) that could otherwise be used for human food and is required to produce 1 kg of animal protein*

Animal	Kg natural protein feed to produce 1 kg animal protein*
Dairy cattle	0.95
Sheep	1.90
Laying hens	2.20
Beef cattle	2.30
Broilers	2.50
Turkeys	4.00
Pigs	5.50

*Based on protein available to humans. See text for further explanation.
(Adapted from T. J. Cunha (1980). The next 20 years in animal production. *Feedstuffs*, 2 June 1980, pp. 45–68. Table 8.)

The need for increased efficiency of animal production
Leaving aside the other important uses of livestock, a good index of efficiency in an animal production system is its ability to provide adequate amounts of affordable food for human consumption. If all American agricultural producers today were using the technology of 30 years ago, it is estimated that the price of food in that country would be two to three times higher than it is. Future production costs will depend similarly on technical advances. There are ethical and political considerations to be taken into account, too: if production costs can be lowered only by 'factory farming' methods, it is for society to decide which methods are justified and how much it is prepared to pay to avoid those that it judges unacceptable.

The means of increasing the efficiency of animal production
The four basic ways of achieving more efficient production are as follows:

1 Improvement of husbandry systems in general, including animal feeding and management, stimulation of growth and lactation, disease control and agricultural economics.
2 Genetic improvements to provide animals that can make best use of a given husbandry system and thus (for example) survive more easily, grow faster, milk better or produce larger litters (discussed in Book 4, Chapter 3, Second Edition).
3 Development of techniques for realizing the full reproductive potential of both males and females by reducing wastage due to failure to conceive, embryonic and fetal death, and mortality during the perinatal period.
4 Reduction of wastage of animal products, particularly in developing countries where preservation and storage methods are often inadequate. FAO figures put food losses (vegetable as well as animal) at 20 per cent world-wide, 40 per cent in South America and 30 per cent in Africa.

Apart from the fourth approach to the problem, the reproductive biologist has an important direct role to play in each. For example, the rate of genetic progress that can be made depends on reproductive technology to proliferate desirable animals and to compensate for the ruthless culling that must be part of any successful selection programme. There is also considerable overlap between the biology of reproduction and the technologies associated with direct stimulation of lactation (as discussed in Book 3, Chapter 8, Second Edition) and artificial control of the growth of animals by hormonal means, including the administration of anabolic steroids, growth hormone or growth-hormone-releasing hormone or, conversely, by immunological neutralization of somatostatin, an inhibitor of growth hormone secretion. These approaches have been succinctly reviewed by Bill Hansel (see Suggested further reading). However, I shall confine myself to a discussion of the many ways in which 'efficiency of animal production' in effect means 'efficiency of animal *reproduction*'.

The room for improvement in the efficiency of animal reproduction
During the 1970s, scientists at the United States Department of Agriculture made some startling estimates of the costs of reproductive inefficiency to animal production in their country, and there is every reason to presume that the figures would be proportionately similar for other Western countries. The estimates are valuable as well as startling because they help define the areas that particularly need attention.

In beef cattle in the USA, probably not more than 80–85 per cent of cows calve, and as many as 5–10 per cent of the calves born die before they are weaned (Table 1.4). It is estimated that merely increasing the net calf crop to 85 per cent could 'save' (as reduced production costs) almost $600 million per year. Proper application of existing technology should be able to boost the crop to about 90 per cent. A further $200 million or more could be saved by reducing the calving interval, so that, on average, cows calved just 6 days earlier in the year. This would reduce maintenance costs for the dams and increase the weights of calves at the end of the grazing season. Thus, there is an immediate potential for reducing beef production costs by about $800 million through increased reproductive efficiency.

In the breeding of American dairy cattle, sources of reproductive inefficiency are about equally divided between failure to detect oestrus on the one hand, and infertility following insemination on the other. The

Table 1.4. *Factors affecting net calf crop in a disease-free beef herd inseminated by natural service (14-year summary)**

Factor	Number of females	Reduction in net calf crop (%)
Females not pregnant at end of breeding season	2232	17.4
Perinatal calf deaths	821	6.4
Calf deaths, birth to weaning	372	2.9
Fetal deaths	295	2.3
Total potential calves lost	3720	29.0
Net calf crop weaned	9107	71.0
Totals	12827	100.0

*Includes females 14 months to 10 years old during mating periods of 45 to 60 days' duration.
(From R. A. Bellows, R. E. Short and R. B. Staigmiller. Research areas in beef cattle reproduction. In *Beltsville Symposia in Agricultural Research*, (*3*) *Animal Reproduction*, pp. 3–18. Ed. H. W. Hawk. Allenheld, Osmun & Co.; Montclair. (1979).)

reasons for the latter are listed in Table 1.5. Reproductive failure is the main reason for the short productive life of the average cow, which lives for only about 5 years and completes only two calvings and lactations. Even in well-managed herds, about 5 per cent of cows fail to conceive during the year and have to be culled. On average, two artificial inseminations are necessary to achieve conception and the time between calvings is in excess of 13.5 months, instead of the ideal interval of 12 months. It is estimated that almost $300 million could be saved by four devices: reducing the average calving interval by 15 days (saving $135 million), reducing calf losses from 10 per cent to 4 per cent ($85 million), reducing the number of 'sterile' culled cows from 5 per cent to 2 per cent ($57 million), and reducing the number of services per conception from 2 to 1.5 ($20 million).

With sheep, a 13 per cent improvement in reproductive efficiency (expressed as weight of lamb marketed per ewe) would save about $8.5 million for the American sheep producer and consumer. There are several tangible ways in which control of reproduction could bring this about, as we shall see.

The efficiency of pig production in the USA remained unimproved during the 1970s, partly because of increased reproductive problems produced by the stress of new intensive management systems. Currently, sows produce an average of 1.8 litters per year with 7.2 pigs from each litter reaching market weight. The potential for improvement is apparent when one considers that, on average, ovulation and fertilization rates for the same sows are 17 and 95 per cent, respectively, but that only 9.4 live pigs are farrowed. If sows could produce two litters per year with 10 pigs marketed per litter, savings would amount to about $500 million per year.

Having established that there is an urgent need to increase the

Table 1.5. *Sources of reproductive failures in dairy cattle*

Cause of failure	Approximate percentage of first inseminations
Anatomical abnormalities	2
Ovulation failure	2
Lost or ruptured ova	5
Fertilization failure	13
Embryonic mortality	15
Fetal mortality	3
Total	40

(From H. W. Hawk. Infertility in dairy cattle. In *Beltsville Symposia in Agricultural Research*, (*3*) *Animal Reproduction*, pp. 19–29. Ed. H. W. Hawk. Allenheld, Osmun & Co.; Montclair (1979).)

reproductive efficiency of our domestic livestock, let us examine ways in which this can be done, first in males and then in females.

Increasing the reproductive capacity of males

This is being achieved through the development of artificial insemination (AI). Two of the fathers of reproductive biology were Cambridge men: Walter Heape performed the first embryo transfer in 1890, and F. H. A. Marshall published the first textbook on *The Physiology of Reproduction* in 1910. They were already aware of the practical importance of AI, which dates back for certain to 1780, and possibly to the 'year 700 of the Hejira'. In 1780, the famous Italian scientist Spallanzani successfully inseminated a bitch and noted:

> ...sixty-two days after the injection of the seed, the bitch brought forth three lively whelps, two male and one female, resembling in colour and shape not the bitch only, but the dog also from which the seed had been taken. Thus did I succeed in fecundating this quadruped; and I can truly say, that I never received greater pleasure upon any occasion, since I first cultivated experimental philosophy.

An even earlier date is suggested in a French anecdote cited by Heape in an 1897 paper on AI. This must also be one of the first papers on pheromones, because Heape describes how an Arab, who owned a valuable mare in oestrus:

> ...armed with a handful of cotton-wool which had been saturated with the discharge from the vagina of the mare, approached by stealth a valuable stallion belonging to a member of a neighbouring hostile tribe, and, having sufficiently excited the animal with the scent of the material he had brought, collected from him spermatozoa, which he introduced on his return into the vagina of his mare, and obtained thereby a foal.

AI was already being used in horse breeding in the USA in the 1890s but, because of stud-book regulations, the practice in horses has advanced only marginally since then. In cattle breeding, on the other hand, it has undoubtedly been the most influential single factor in augmenting productivity. AI of cattle began to be used extensively in Russia and Japan at the beginning of this century. In Denmark it was first used in 1936, in the USA in 1937 and in Britain in 1942. Today, its use in dairy cows varies from almost 100 per cent in Denmark and Japan, through about 60 per cent in the USA and down to much lower percentages in less developed countries. Dairy cows lend themselves particularly well to AI because their intensive management makes heat detection and physical handling at the time of insemination relatively easy. These factors pose more difficulties in extensively managed beef cattle, so beef producers in North America make less use of AI; only about 4 per cent of beef cattle are artificially inseminated. However, in countries like Japan, where beef is produced in intensive systems, AI is more popular.

The use of AI in pigs is increasing substantially and is estimated to involve between 5 and 10 million animals annually, the USSR probably being the largest user. AI is now used for over 30 per cent of pig services in the Netherlands, Norway, Finland, East Germany and the Bavarian province of West Germany. Sheep AI is not yet very widespread, but it is important in France, the USSR and China, and the success rate is improving as a result of uterine insemination performed with the aid of laparoscopy.

Let us now consider the basic principles and procedures of AI, and examine their impact on animal production. The bull will be taken as the example and males of the other domestic species will be compared with him.

Principles and procedures of AI
Male selection and management. Proper assessment of the genetic worth of males used for AI is of extreme importance because of the very large numbers of offspring that each male can sire. Popular bulls, for example, have produced as many as 50 000 calves in a year, although this number is more usually halved with today's use of frozen semen. Since frozen semen can be used long after a bull's death, total numbers of offspring can be as high as 150 000.

Selection is usually a two-stage process. Young bulls, themselves the progeny of carefully planned contract breedings, are used to provide semen for about 400 inseminations soon after puberty, when about 12 months of age. This allows at least 50 or 60 of their progeny to be evaluated in performance tests. These tests are usually of the milk production of their daughters, or the growth rates and meat characteristics of their male and female offspring, depending on whether the bulls are of dairy or beef breeds. Once the results of the progeny tests are available (in 4–5 years), only the very best bulls are put into service at the AI stud. The others are slaughtered, because severe culling is essential to maintain genetic progress. Physical and, in the case of beef breeds, performance characteristics of the bulls themselves are also taken into account in making selections. For example, testis size has been shown to be highly correlated with sperm production, is highly heritable, and may even be correlated with ovulation rates and the early attainment of puberty in related females. Cytogenetic screening is practised in some countries to avoid transmission of defects such as chromosome translocations which can reduce fertility. This has assumed increased importance recently as the relationship between the 1/29 translocation and infertility becomes more firmly established and as the success rate of selection against the translocation improves (since 1969 the incidence of the defect has been considerably reduced from an initial level of over 12 per cent in the Red and White breed in Sweden, with a concomitant increase in fertility). A further criterion for male (especially ram) selection is the survival rate of their spermatozoa after freezing and thawing, because this vital quality varies widely between individuals.

The large bull studs associated with today's AI organizations are designed around central semen collection and processing facilities. The closest attention is paid to quarantining bulls to prevent them contracting and disseminating sexually transmitted diseases. Prevention of disease is also one of the reasons why the United Kingdom, for example, now prohibits the use of fresh semen; the use of frozen semen can be delayed until it is certain that the bull was not incubating an infection at the time of collection.

Semen collection and processing. Routine collection methods for most of the domestic species depend on inducing the male to ejaculate into an artificial vagina (Fig. 1.1), and collecting the semen in an attached container which is insulated to prevent damage to the spermatozoa from sudden temperature changes. With boars, the more usual method is to grasp the spiral end of the penis with a gloved hand, the fingers simulating the female's cervical ridges which interlock with the glans penis during normal mating. Electro-ejaculation (direct stimulation of the ampullae and seminal vesicles by a square-wave electrical pulse via a rectal probe) is confined to rams and goats, and to young, inexperienced or physically incapacitated bulls.

Several aspects of sexual behaviour are of practical importance in semen collection. First, males of all the domestic species can be trained to serve an artificial vagina, and will mount live 'teaser' animals (whether or not in oestrus) or dummies to do so. Secondly, sexual preparation of the bulls (e.g. by walking around the teaser, allowing preliminary false mounts, introducing a second bull, or changing the location of the teaser) improves semen quality by increasing numbers of spermatozoa by up to 100 per cent. Similar measures do not seem to improve a stallion's semen. Thirdly, individual bulls' preferences have to be taken into account, relating, for example, to the temperature of the artificial vagina or the texture of its rubber liner. Fourthly, species differences in speed of ejaculation have to be appreciated; these range from seconds in sheep and goats to several minutes in boars. Skilled stockmen who understand these facts can have a very considerable influence on the males' semen production, and thus on the full exploitation of their reproductive potential.

Bulls normally produce 4–10 ml semen per ejaculate and samples can be collected two to six times per week. Rams and billy-goats produce only 0.5–1.5 ml, but collections can be repeated up to 25 times per week. Boars and stallions produce large volumes of gel from the bulbo-urethral (Cowper's) glands in addition to large volumes of gel-free semen (150–300 ml for boars; 30–100 ml for stallions), but even so, collections can be made as often as from bulls. After collection and evaluation of the semen for sperm density and motility, ejaculates can be 'extended' (diluted) to varying degrees with a variety of buffered isotonic solutions containing a carbohydrate energy source (glucose or fructose), protective proteins (usually in the form of milk and/or egg yolk), antibiotics, and, for semen that is to be frozen, cryoprotective agents (glycerol and/or dimethylsulphoxide). Bull semen can be extended 10–75 times, ram semen 5–10 times and goat semen 10–25 times, but boar and stallion semen only 2–4 times. The volume of frozen–thawed semen necessary to inseminate cows is 0.2–1.0 ml (10–15 million motile spermatozoa), ewes 0.05–0.2 ml (200 million), goats 0.5 ml (200 million), sows 50 ml (5000 million), and mares 20–50 ml (1500 million). These figures determine how many females can be inseminated with the semen from a single male in a given time, which

Fig. 1.1. Semen collection from a stud bull at Canada's largest AI centre, Le Centre d'Insémination Artificielle du Québec (CIAQ), Saint-Hyacinthe, Quebec. (*a*) The bull's penis is deflected into the artificial vagina (AV) as he mounts the teaser animal. (*b*) The conical semen receptacle, still within a warm-water jacket and attached by its latex sleeve to the AV, after removal from its insulating cover (seen in the background).

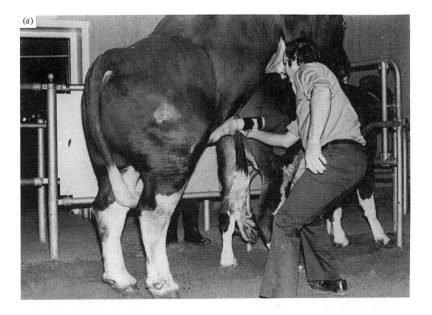

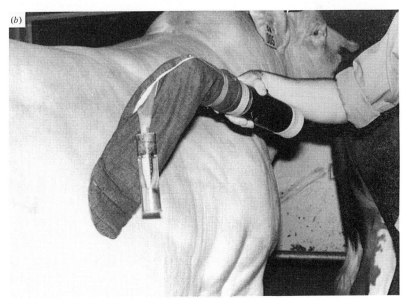

ultimately determines a male's genetic influence (see Table 1.6 and also Book 4, Second Edition, Fig. 3.14).

Even with the very best management and semen collection techniques, it is only possible to obtain just over 7000 million spermatozoa per day from bulls. Since the testes of a mature dairy bull produce about 13 000 million spermatozoa per day, about 50 per cent must be lost through resorption in the epididymis, or wash-out from the pelvic urethra at urination, or masturbation.

The discovery that glycerol will protect spermatozoa from the lethal effects of freezing really qualifies for the overused accolade of being a 'breakthrough'. It was made in 1949 by Chris Polge and colleagues working at Mill Hill, London, England, and owes much to serendipity. They were testing the effects of various sugars borrowed from the biochemistry department and suddenly found an effective one in an old bottle mislabelled 'fructose'. The contents turned out to be Meyer's egg-albumen, which contains glycerol! Frozen bull semen was originally stored in glass ampoules or as pellets made by dropping it directly on to 'dry ice'. Now, most is frozen, stored, thawed and inseminated in plastic straws (such as the types made in Germany or by the French company Cassou), developed from real straws first used in early Danish, Russian and Japanese work with fresh semen. Semen processing in plastic straws is now highly automated, and millions of doses of bull semen are stored in large tanks of liquid nitrogen in AI centres throughout the world (Fig. 1.2).

AI with frozen semen of the other farm animals has not been as well

Table 1.6. *Average numbers of females that can be inseminated and conception rates that can be obtained with frozen–thawed extended semen of domestic animals*

Animal	No. of females that can be inseminated		Conception rate to first insemination (% pregnant)
	Per ejaculate	Per week	
Cattle	300	1000	60
Sheep and goats	15	150	50 (sheep)
			60 (goats)
Pigs	10*	30*	65
Horses	5†	15†	30–50

* Two inseminations during oestrus are recommended for sows.
† Repeated inseminations on alternate days are necessary during the mare's prolonged oestrus.
(Adapted from R. H. Foote. Artificial insemination. In *Reproduction in Farm Animals*, 4th ed., pp. 521–45. Ed. E. S. E. Hafez. Lea & Febiger; Philadelphia (1980).)

developed, partly because there has been less commercial incentive to make it work. However, the use of frozen semen in controlled sheep breeding programmes is now spreading. Freezing does have a deleterious effect on semen, even from bulls, and so dilution cannot be as extensive as when it is used fresh. Although this limitation is usually outweighed by the advantages of being able to stockpile frozen semen, there are circumstances (in New Zealand, for example) where fresh semen is still used to some extent so that larger numbers of cows can be inseminated from a given bull in a short breeding season.

Fig. 1.2. Semen storage in liquid nitrogen at Le Centre d'Insémination Artificielle du Québec (CIAQ), Saint-Hyacinthe, Quebec. This room contains enough semen, frozen in straws, to inseminate three million cows.

Fig. 1.3. For the artificial insemination of a cow, the cervix is held through the wall of the rectum and straightened, and a pipette (or an AI gun holding a straw) is passed into the uterus for deposition of the semen just within the body of the uterus and/or the most anterior folds of the cervix. (After G. W. Trimberger. In *Reproduction in Farm Animals*, 1st ed., fig. 8–5. Ed. E. S. E. Hafez. Lea & Febiger; Philadelphia (1962).)

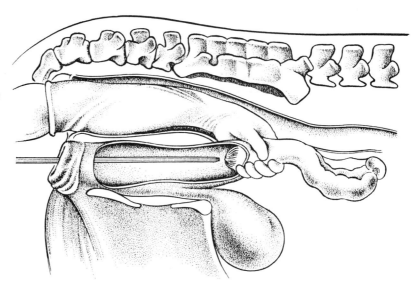

Semen thawing and insemination. Bull semen can be thawed at a variety of controlled temperatures and inseminated directly into the cervix and body of the uterus (Fig. 1.3). For maximum fertility, insemination should closely precede ovulation. In cattle, this means inseminating during the latter half of oestrus or at a predetermined time after treatment to synchronize oestrus. Such timing is important for ensuring that viable spermatozoa are present and ready to fertilize eggs as soon as they are ovulated, because the fertilizable life of an egg is relatively short. The practical effects of inseminating pigs at various times relative to ovulation illustrates this point even with fresh semen (Fig. 1.4).

Fig. 1.4. Conception rate and litter size in pigs artificially inseminated at different times relative to the onset of oestrus, and the estimated time of ovulation. (Drawn from data of J. Boender. *World Review of Animal Production*, Special issue 2, p. 29 (1966).)

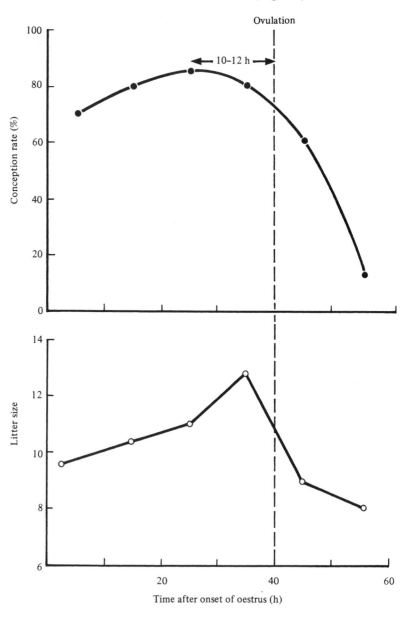

Thawed semen is expelled directly from straws with a specially designed 'gun' so that there is no wastage (Fig. 1.5). When pelleted semen is used (notably with sheep and pigs), is has to be thawed in a suitable extender. In sheep, it is usually centrifuged to concentrate the spermatozoa into a small volume (less than 0.2 ml) for intracervical deposition through a vaginal speculum (a difficult procedure), or intra-uterine injection at laparotomy or laparoscopy, which can be surprisingly rapid in experienced hands. In contrast, sows are inseminated with 50 ml of the extended semen by an insemination pipette that locks into the cervix by a cuff or a spiral thread simulating the boar's penis (Fig. 1.6). This large volume is necessary to facilitate sperm transport through the 3-metre-long uterine horns (see Chapter 2, Book 4, Second Edition). Mares also need a large volume (20–50 ml) of inseminate, injected by syringe through a pipette, which can be readily guided through the cervix with a finger. Repeated inseminations are recommended in sows and mares (see Table 1.6) and producers of pigs for slaughter are making increasing use of pooled semen for AI, or are combining AI and natural service, in order to improve fertility.

Assessment of semen. Although experienced personnel can assess the normality of semen by simple microscopic examination, there is still no reliable laboratory test that can accurately predict the fertilizing capacity of a given sample. It remains to be seen whether tests based on sperm

Fig. 1.5. A Cassou AI gun. (*a*) and (*b*) show the proximal and distal ends, respectively, of its component parts. These are: a steel plunger with a plastic hub (A); a 0.5 ml plastic straw with a white plug of polyvinyl alcohol between two lengths of wick (B); a stainless steel barrel with a plastic hub (C); a plastic locking ring (D); and a clear plastic pipette with a tapered tip (E).

The straws would be purchased as a sealed unit containing semen and stored in liquid nitrogen. After carefully controlled thawing, the heat-sealed tip is cut off with scissors and the straw is fitted into the distal end of the barrel. The clear plastic pipette holds it there by enclosing the barrel; it is locked to the hub of the barrel by the ring. The plunger expels the semen from the straw when the plug is pushed through to its distal end. (*c*) The assembled gun with the plunger inserted as far as the proximal end of the straw, ready for insemination. (*d*) The distal end of the gun after discharge. The plug can be seen pushed to the end of the straw that protrudes from the barrel inside the plastic pipette. (Photographs courtesy of Robert Bériault.)

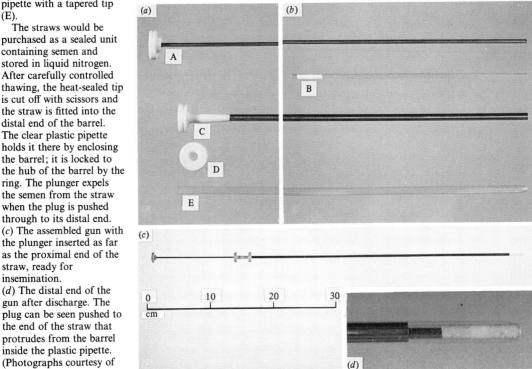

penetration of cervical mucus or of zona-free hamster oocytes live up to their early promise and remedy this situation, as they have to some extent in human medicine. The spermatozoa from two different bulls can be compared by mixing equal numbers of spermatozoa from both bulls for 'heterospermic insemination', and then determining which bull sires the

Fig. 1.6. Diagrams from an instruction manual of pig AI (issued by the Meat and Livestock Commission Pig Breeding Centre in England) showing how the spiral insemination catheter should be twisted to lock it into the cervix, and how semen is then slowly introduced through the catheter by squeezing a collapsible plastic bottle over a 5–10 min. period.

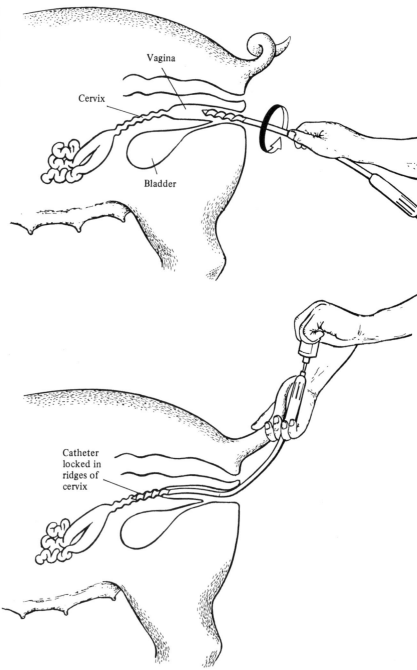

more offspring. Such a test is obviously time-consuming, but can be shortened by counting early embryos after spermatozoa from one of the bulls have been marked with a fluorescent dye, or treated with an alkylating agent such as tris-(1-aziridinyl)-phosphine oxide (TEPA) which arrests egg development after one or two divisions. Semen samples must be compared over a range of dilutions because differences may be masked until one of the samples is used at threshold concentrations. Similarly, insemination should be carried out at varying times before ovulation, to show up possible differences between bulls in the longevity of their spermatozoa in the female tract.

The impact of AI

Robert Foote of Cornell University has neatly summarized the way in which reproductive physiology interacts with genetics in AI by two simple equations:

$$\text{Genetic impact of sire} = \text{Genetic superiority of sire} \times \text{Number of progeny per sire} \qquad (1.1)$$

$$\text{Number of progeny per sire} = \frac{\text{No. of sperm produced/sire}}{\text{No. of sperm required/female}} \times \text{Fertility of the semen} \times \text{Fraction of the spermatozoa used for insemination} \qquad (1.2)$$

For Equation 1.2 with bulls, Foote assumes:

1 Total spermatozoa harvested per sire per year $= 1.5 \times 10^{12}$.
2 Spermatozoa required per cow $= 15 \times 10^6$ (although many AI organizations use more).
3 Fertility $= 50$ per cent (0.5).
4 Semen used $= 100$ per cent.

Solving Equation 1.2: annual number of progeny per sire $= 50\,000$, which, in Equation 1.1, can be seen to represent a very considerable amplification of any genetic superiority of a given bull. Detection of that superiority should improve as the proportion of progeny-tested bulls retained for extensive use declines; improving on 50 000 progeny is very much up to reproductive biologists.

The impact of genetic improvement on milk yield over 20 years in the USA is depicted in Fig. 1.7. While it represents only about 30 per cent of the total improvement, it should be noted that it contrasts with environmental improvements in that it is permanent. By 1984, as much milk (some 5500 million kg per year) was being produced from just under 10 million American cows as was produced by 27 million cows in 1950. The development of today's effective tests for detecting genetic superiority has been due largely to the incentive provided by the AI industry. However, there is still much room for improvement in these tests because genetic progress to date has been less than that theoretically possible. This is probably because some form of bias exists in our methods for selecting bulls' parents, and because selection itself has not been as rigorous as it should be once test results are known. It must also be remembered that selection for one trait, such as milk yield, cannot be made entirely independently of selection for other characteristics such as conformation.

Apart from genetic improvements, other benefits of AI include the elimination of venereal diseases from the inseminated herd, and the fact that it is no longer necessary to keep dangerous bulls on the farm.

Increasing the reproductive capacity of females

Reproductive efficiency in females is obviously essential for maximum production in minimum time, and also to realize the full potential for genetic improvement offered by widespread application of AI. Several aspects of female reproductive physiology are subject to varying degrees of control in the following procedures:

1 Detection of oestrus.
2 Induction of oestrus and ovulation prepubertally or outside the normal breeding season, and increased ovulation rates.
3 Synchronization of oestrus and ovulation in groups of animals.
4 Embryo transfer and related techniques of embryo preservation and manipulation.
5 Diagnosis of pregnancy.

Fig. 1.7. Factors affecting improvement of milk yield over 20 years in the USA. The production recorded was in 2-year-old Holstein cows. (After R. H. Foote. In *New Technologies in Animal Breeding*, p. 23, fig. 3. Ed. B. G. Brackett, G. E. Seidel and S. M. Seidel. Academic Press; New York (1981).)

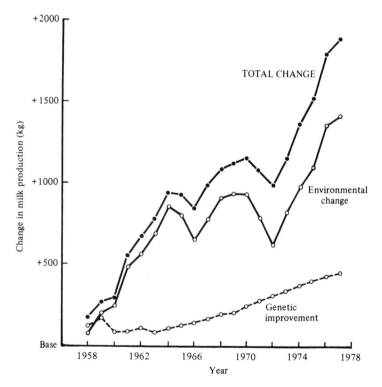

6 Control of the timing of parturition and reduction in perinatal mortality.

7 Reduction in embryonic and fetal mortality.

Every one of these procedures owes something to research in basic reproductive biology, especially the endocrinological work of the past 25 years. Many of the procedures depend on the availability of hormones and their analogues developed by pharmaceutical companies. Unfortunately, these companies no longer have much incentive to continue working in this field, for two interrelated reasons: first, the limitations of agricultural economics; second, the enormous expense of proving to the drug regulatory agencies that any new drug that could enter the human food chain is 'safe' to market.

Detection of oestrus

Oestrus detection is the most fundamental of a stockman's skills. It is surely an instructive paradox that poor oestrus detection should be the primary reason for poor AI results in modern, sophisticated dairy herds and a major problem in large intensive piggeries. The scale of the problem is directly related to the increase in intensive management. This results, for example, both in decreased time for herdsmen to observe the behaviour of their cows in highly automated dairies, and in husbandry systems (such as sow tethering) that reduce the animals' opportunities to show signs of oestrus. Estimates from various countries agree that about half of all periods of oestrus pass undetected on many dairy farms. Conversely, progesterone determinations in blood or milk have shown that about 20 per cent of cows presented for insemination are not, in fact, in oestrus. Small wonder, then, that recent studies of 25 000 artificially inseminated cattle in New York and California have revealed calving rates of only 50 per cent and 44 per cent, respectively; in the Californian study, the contemporary calving rate in naturally mated cows was 63 per cent. This suggests that there must be a point at which labour saving becomes a false economy. The problem is so severe that some producers are giving up AI and returning to natural service, although genetically this is a highly retrograde step. As it is most unlikely that labour costs will go down, we should examine what biological facts can be useful in improving oestrus detection.

First and foremost it must be emphasized that the outward signs of oestrus may be brief and need to be understood. They may last no longer than 2 h and are often apparent for less than 7 h, and are usually at night, as has been confirmed by continuous videotaping. Mere appreciation of this may encourage a rethinking of observation patterns for oestrus detection, and should discourage the notion that cattle are subject to 'silent' ovulation. Because cattle exhibit homosexual mounting behaviour, males are not essential for oestrus detection. Nevertheless, some producers prefer to use steers (castrated males), or heifers treated with androgens, or entire bulls to detect oestrus, equipping the 'teaser' with an oversized ball-point pen on its chin halter to mark the animals that it mounts (Fig. 1.8). Bulls have to be subjected to somewhat questionable surgery involving deflecting the penis to one side, thereby preventing unwanted insemination and the spread of venereal disease. Detection of mounting can be helped by glueing devices to, or painting, the cow's sacral area. The sternum of the animal mounting the one in oestrus will scrape off the paint or cause the pressure-sensitive device to change colour.

Changes in body temperature are of no use for predicting ovulation in cattle. A decrease in the electrical resistance of the vaginal mucosa occurs at oestrus, when vaginal mucus production is maximal. Electronic probes to detect this are on the market, though their practical usefulness on a field scale remains to be proven. Of considerable biological interest is the fact that dogs can be trained to identify oestrous cows by detecting vaginal odours, raising the possibility that the odoriferous substance(s) will be identified and one day detected chemically.

Detection of oestrus in sheep and goats relies on marking by teaser males equipped with a marker crayon or paint on the brisket. In mares, it is necessary to observe the behavioural response to the presence of a stallion; when in heat the mare will urinate and erect the clitoris ('winking'). Oestrus in sows can be detected in the absence of the male by evoking a reflex immobility by pressing or sitting on their backs. This is helped by spraying the sow's snout with the pheromonal steroids that give the boar his odour (see Book 3, Chapter 6, Second Edition). The difficulties of detecting oestrus in intensively managed systems for all species have been an incentive to control the timing of oestrus and ovulation so as to permit insemination to be performed at fixed times after treatment.

Fig. 1.8. A 'bull-point' chin marker used on teaser bulls, steers or androgenized heifers or cows to detect oestrus. Ink from the device is left on the shoulder region of females that have been properly mounted.

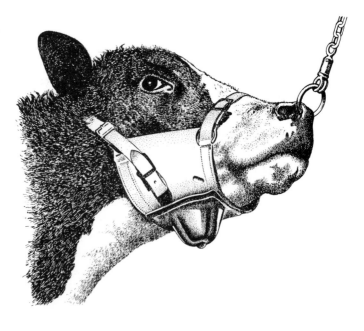

Induction of oestrus and ovulation

The rate of reproduction could be increased if puberty could be accelerated, or if ovulation could be induced during lactational anoestrus in cattle and pigs, or during seasonal anoestrus in sheep, goats and horses. There is a limit to the extent to which genetic progress can be accelerated by breeding from younger animals; it would not make sense to induce pregnancy in females too immature to carry young and give birth. However, there are some situations in which it could make sense to hasten puberty. For example, seasonally breeding animals like sheep, goats and horses may lose almost a year of reproductive potential by just failing to reach puberty in time for the mating season. Although not really seasonal breeders, cattle also need to be mated at certain times of the year to fit in with management plans, particularly in beef production. Delayed puberty is becoming an increasingly serious problem amongst intensively housed pigs, probably because of ill-defined stresses. Specially valuable females need not necessarily complete their pregnancies, but could be put to better use by providing embryos for transfer to recipients, or oocytes for *in vitro* fertilization and transfer.

The basic principle of ovulation induction in any of these circumstances is to provide a sufficiently mature ovarian follicle with sufficient gonadotrophins to induce it to rupture and release a mature oocyte. The gonadotrophins can be either exogenous or endogenous. Exogenous gonadotrophins may be prepared from extracts of pituitaries from slaughtered animals (FSH and LH) or placental hormones derived from blood or urine. The one from blood is equine chorionic gonadotrophin (eCG), more commonly (but misleadingly) called pregnant mare serum gonadotrophin (PMSG). The one from urine is human chorionic gonadotrophin (hCG). (The properties of these hormones are discussed in Book 7 of the First Edition and in Book 3 of the Second Edition.) Endogenous gonadotrophins can be released from the pituitary by hypothalamic gonadotrophin-releasing hormone (GnRH) which, in its turn, can be exogenous or endogenous. Exogenous GnRH can be injected intermittently or continuously. Endogenous GnRH release can be stimulated in a number of ways: by oestrogen treatment; by manipulation of the photoperiod by melatonin administration in sheep, goats and mares; by removal of the inhibitory effect of suckling in cattle and pigs; by blocking the feedback inhibition of gonadal steroids on the hypothalamus with antagonists or antibodies.

Species differences often dictate the type of treatment to be used. Pigs come into oestrus when their follicles are stimulated by exogenous gonadotrophins, and so satisfactory conception rates can be achieved by insemination of prepubertal gilts treated with eCG and hCG, or eCG and GnRH. Pregnancy is maintained at acceptable levels only in gilts that are close to puberty because mechanisms for maintaining corpora lutea are deficient in younger ones. Physiological doses of oestradiol benzoate are also showing promise for inducing puberty in gilts after 160 days of age. These are much more effective in winter than in summer, suggesting underlying seasonal patterns of reproduction in pigs.

Anoestrous sheep and goats may be induced to ovulate in response to gonadotrophin therapy, but unlike pigs, they do not come into oestrus; it seems that oestrous behaviour depends on prior conditioning of neural centres with progesterone. In order to induce oestrous behaviour, therefore, it is first necessary to simulate a luteal phase by 1–2 weeks of progestogen treatment. This is administered by one of a variety of routes, depending on the pharmacology of the compound and the animal management system. The progestogen may be injected (frequently, or once in a slow-release formulation), given orally in food or water, implanted subcutaneously in material from which it slowly diffuses (such as silicone rubber), or placed in the vagina, either in impregnated sponges or in the silicone rubber covering of a spiral metal band known as a 'progesterone-releasing intravaginal device' (PRID, see Fig. 1.9). The progestogen behaves like endogenous progesterone during the luteal phase of the oestrous cycle and suppresses endogenous gonadotrophin release. To simulate normal luteolysis, the treatment is abruptly terminated by

Fig. 1.9. A progesterone-releasing intravaginal device (PRID) for use in cattle, with an explanatory diagram from its manufacturer's (Abbott Laboratories) instruction sheet. Crystalline progesterone is embedded throughout the silicone rubber coating of the spiral metal band. For insertion, the spiral is tightened to a smaller diameter inside a vaginal speculum. When pushed out from the speculum, it resumes its larger diameter and clings to the vaginal wall. The cords are left hanging from the vulva and are used to withdraw the PRID when required. Oestradiol benzoate contained in the gelatin capsule produces a short-lived peak of oestradiol-17β at the time of insertion. This is said to cause luteolysis, so that no endogenous progesterone complicates the pattern of progesterone decline when the PRID is removed. However, many users find the capsule unnecessary and remove it before PRID insertion (see Table 1.7).

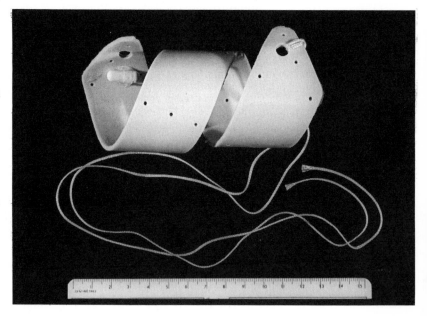

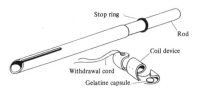

stopping the injections or feeding, or by withdrawing the subcutaneous or vaginal implants. Just before this, the animal is given an eCG injection to induce follicular development and oestrogen secretion. Oestrogen (either endogenous or injected) brings about an LH surge, behavioural oestrus, and ovulation. Therapy of this kind is being widely used for out-of-season sheep and goat breeding in Europe, but has yet to find much application elsewhere. It must be remembered that male sheep and goats are also seasonal breeders, and so, unless AI with stored frozen semen is used, they must be in breeding condition when the females are induced to come into oestrus and ovulate.

Photoperiodic control depends on light-proof housing for sheep because the onset of the breeding season is induced by diminishing day-length. Such treatments are appropriate only for intensive management systems. In horses, on the other hand, it is increasing daylight in the spring that stimulates reproduction, and this only requires extra lights and time-switches in the loose box.

There are now exciting developments in the search for more practical alternatives to the physical control of the photoperiod for controlling seasonal breeding. They are based on treating animals with melatonin, which is normally released from the pineal gland during the hours of darkness, and thereby enables the animal to sense the seasonal changes in daylight length (see Book 3, Chapter 3, Second Edition). In all mammalian species studied so far, increasing the number of hours in the day during which circulating melatonin concentrations are elevated induces the short day-length response normally seen in autumn and winter. Since melatonin is now readily available as a by-product of decaffeinated coffee production, and is effective when given orally (absorbed into a pelleted diet), or as a subcutaneous implant or injection, it has become a practical means of treating farm animals. Thus, ewes can be fed the hormone late in the afternoon when in summer anoestrus, thereby mimicking an autumnal melatonin secretion which, in turn, stimulates the onset of the breeding season. Unfortunately, although this is a convenient way of fooling animals into behaving as though it were winter during the summer, there is not yet a simple way of accomplishing the reverse. It would be highly desirable to be able to do so, not only to stimulate winter mating in long-day breeders such as horses, but also to increase productivity by reversing the winter anorexia usually seen in farm animals (see also Book 4, Chapter 4, Second Edition). Reducing effective melatonin levels to bring this about will require specific anti-melatonin compounds which could act either by blocking its release from the pineal, or by neutralizing it in the circulation, or by blocking its action on the brain.

Much research has been directed at controlling lactational anoestrus in cattle, which contributes significantly to prolonged calving intervals. Attempts to initiate cyclic activity post-partum with GnRH have met with variable success. Early weaning or restricted suckling shortens the duration of post-partum anoestrus by releasing the pituitary from suckling inhibition (see Book 3, Chapter 6, Second Edition). Even removing the calf for 48 hours appears to help the oestrus- and ovulation-inducing effect of a 9-day progestogen treatment in beef cows. Post-partum anoestrus in sows can also be shortened by early weaning; sows also respond like prepubertal pigs to eCG–hCG therapy, the response improving as lactation progresses. Gonadotrophin treatment is therefore being used when management systems inhibit the anovulatory oestrus that normally occurs 4–5 days after weaning. Oestradiol benzoate injections at critical times after weaning are also showing promise for synchronizing oestrus and ovulation in sows.

The hypothalamus–pituitary–ovarian axis (see Book 3, Chapters 1 and 5, Second Edition) has been exploited in quite another way in order to increase ovulation rates in sheep, with about a 20 per cent increase in lamb production. The method takes advantage of observations first made by Rex Scaramuzzi and colleagues in Edinburgh in the 1970s, when they showed that active immunization of ewes against various ovarian steroids, but particularly androstenedione, could reduce their feedback inhibition of gonadotrophin production. The consequent increase in gonadotrophin levels led to moderate increases in ovulation rate (by 18–45 per cent) without affecting oestrous behaviour. A commercial preparation of androstenedione conjugated to human serum albumin, together with DEAE dextran, is now widely and successfully used in a two-injection active immunization schedule just before introducing the rams. This ensures high titres of the neutralizing antibody during the mating season. The titre declines slowly thereafter so that the procedure must be repeated before the next breeding season.

Another immunological means of increasing ovulation rates and advancing puberty in sheep has resulted in Australia from studies of the Booroola Merino which, on average, spontaneously produces just over four oocytes at each oestrus (Book 4, Chapter 3, Second Edition). Bernie Bindon and his colleagues have shown that Booroola ewes are very deficient in the ovarian follicular hormone inhibin which normally keeps FSH secretion in check by 'negative feedback' on the anterior pituitary (Book 3, Chapter 5, Second Edition). Consequently, circulating FSH levels in Booroola ewes are higher than in normal Merinos. With a view to reproducing this situation artificially, in 1982 the CSIRO group began immunizing normal lambs against bovine follicular fluid (as a crude source of inhibin) and has produced sheep that not only reach puberty at a younger age than usual but also regularly ovulate 6 to 8 oocytes. Inhibin has since been isolated from granulosa cells, its structure defined and its gene cloned by DNA technology, so the prospects for using refined versions of this approach to increase ovulation rates in sheep (and probably other species) are bright indeed.

Synchronization of oestrus and ovulation

Inducing oestrus synchronously at predetermined times in groups of animals helps the farmer by limiting the period during which the females have to be observed and handled both at mating and at parturition and is ideal for artificial insemination since it makes the most of the inseminator's visit. Both factors reduce the costs of AI and should thereby encourage its spread, particularly in beef herds and among dairy heifers, where it is at present underutilized. Embryo-transfer donors and recipients must also be brought into oestrus synchronously and this is often accomplished by hormone treatment. There are two basic approaches to the synchronization of oestrus and ovulation:

1 Progesterone or synthetic progestogen treatment is used to prolong the luteal phase of the cycle, and this is followed by withdrawal of treatment to allow synchronous resumption of the follicular phase.
2 'Luteolytic' agents such as prostaglandin $F_{2\alpha}$ are used to lyse (destroy) corpora lutea, allowing synchronous resumption of the follicular phase.

Each approach has numerous variants and many practical systems now make use of aspects of both of these procedures.

The initial demonstrations that repeated progesterone injections inhibit oestrus and ovulation in cattle and sheep were made between 1948 and 1955. Bill Hansel, whose group at Cornell University has added enormously to our knowledge of the cow's oestrous cycle and its artificial control, has divided the history of oestrus synchronization since then into four overlapping phases.

Phase I began in the 1950s when, as a by-product of human contraceptive development, orally active progestogens became available; around 1960 it was found that, when administered by any of the routes referred to in the previous section, these compounds could synchronize oestrus quite effectively in 80–90 per cent of cattle and sheep over a 4-day period, starting 2 days after withdrawal. The progestogens were of no use in pigs because they caused cystic follicles to develop in the ovaries. By about 1967, enough work had been done to show that conception rates in synchronized cattle and sheep were at least 15 per cent lower than in control (untreated) groups. Slow-release implants of progesterone itself were shown to increase the effectiveness of synchronization in cattle as the treatment period was extended up to 21 days. Unfortunately, the conception rate declined after extended treatments, but was as good as in untreated controls after short treatments (9–12 days). This finding explains the rationale of the next developmental phase.

Phase II was associated with efforts to conserve the high fertility that followed short-term progestogen treatment and, at the same time, to improve the synchronization by additional injections of either oestrogen or gonadotrophins. Oestrogen given near the beginning of progestogen treatment causes luteolysis, so that endogenous progesterone production does not subsequently complicate the decline in progestogen levels at the time of implant withdrawal. Such combinations did, in fact, improve synchrony and fertility. Injecting gonadotrophin (eCG or hCG) at the end of the synchronizing treatment was ineffective in cattle but very useful in sheep.

Phase III began in 1972 with reports of the luteolytic effects of prostaglandin $F_{2\alpha}$ in cattle. Work throughout the early 1970s confirmed the effectiveness of this compound and its synthetic analogues (PGs) for inducing an ovulatory oestrus 3–4 days after injection, providing the corpus luteum is at least 5 days old when the PG is given. The ineffectiveness of PGs during the first 5 days after ovulation or in the absence of a corpus luteum (more than 16 days after ovulation) must be taken into account when synchronizing groups of animals. Injections can either be confined to animals in which a mature corpus luteum can be palpated (which involves skilled manpower and animal handling) or given to all animals on two occasions with an 11-day interval. The double injection regimen ensures that animals 0–5 or 16–20 days after ovulation at the first injection will be 11–16 or 6–10 days after ovulation at the time of the second treatment. Insemination may be carried out at fixed times after treatment, or at the time of detected oestrus. There are innumerable variations on these basic themes. Conception rates can be as high as 70 per cent provided that other management factors are good, but they seem more variable after the two-injection treatment. It is interesting that, in the United Kingdom at least, more PG is used to treat individual dairy cows to maintain proper calving intervals than is used for group synchronization, despite the fact that conception rates in lactating cows treated with PG have generally been less satisfactory than in heifers.

Phase IV covers the current attempts to combine progestogen and PG treatments. Experimentally, a 7-day treatment of Holstein heifers with a progesterone-releasing intravaginal device, coupled with PG on the sixth day, has resulted in improved synchronization (73 per cent in oestrus within a 24 h period) and a satisfactory 66 per cent conception rate (Table 1.7).

Research continues into methods of synchronization, and opinions vary as to the best method currently available. The fact is that adequate technology for efficient control of the oestrous cycle in cattle, sheep and goats already exists; it is its application in the field that lags behind. Complicating factors in some countries are government licensing regulations. In the USA for example, the use of PG has only recently been permitted, and no synthetic progestogens suitable for sheep synchronization are on the market. Methallibure, a very satisfactory alternative to synthetic progestogens for blocking pituitary release of gonadotrophins in pigs, had to be withdrawn from the market when it was found to be teratogenic. PGs are not of practical use in cyclic pigs because the corpora lutea remain insensitive to them for 12 days after ovulation.

Synchronization of oestrus and ovulation in mares is made more difficult by peculiar reproductive phenomena such as persistent corpora lutea, ovulations during the luteal phase and, especially, by variability in the duration of oestrus. PG, for example, usually synchronizes the onset of oestrus quite effectively, but a mare with a 4-day heat will then proceed to ovulate well before one with a 7-day heat. This can be obviated by inducing ovulation with hCG, but opinions vary as to whether the hormone can be used repeatedly or whether it becomes ineffective because of antibodies formed against it. Fairly effective regimens of progestogen–PG–hCG/GnRH, or repeated injections of an oestrogen–progesterone mixture, have been worked out, and a possible new approach involves exogenous FSH treatment during the follicular phase of the cycle. The mare is especially sensitive to $PGF_{2\alpha}$; relatively low doses produce luteolysis and special analogues are available that avoid adverse side-effects such as colic and sweating.

Embryo transfer and related embryo manipulations
In all the farm mammals, it is possible to remove unattached embryos from the uterine lumen of their mother (the donor) and transfer them to the uterus of other females (recipients) for development to term. The first embryo transfer was performed by Walter Heape in rabbits as long ago as 1890, and successes in farm animals date from the 1930s for sheep and goats, the early 1950s for pigs and cattle, and the 1970s for horses.

Table 1.7. *Pregnancy rates in Holstein heifers after a single insemination at a preset time after oestrous-cycle control with prostaglandin $F_{2\alpha}$ (PG) alone or in combination with a progesterone-releasing intravaginal device (PRID)*

Treatment	Animals treated	Pregnant (%)*
Control	91	73[a]
2 × PG	90	52[b]
PRID-7 + PG6	93	66[a]

* The 2 × PG treated heifers were inseminated at 80 h after the second PG injection. The PRID-7+PG6 heifers were inseminated at 84 h after PG injection. Controls were inseminated at the first observed oestrus during a 25-day breeding period.
[a, b] Values with different superscripts differ ($P < 0.05$).
(From R. D. Smith, A. J. Pomerantz, W. E. Beal, J. P. McCann, T. E. Pilbeam and W. Hansel. Insemination of Holstein heifers at a preset time after estrous cycle synchronization using progesterone and prostaglandin. *J. Anim. Sci.* **58**, 792–800 (1984). In these studies the oestradiol benzoate capsule was removed from the PRIDs.)

Fourteen years ago, in the first (1972) edition of this book, embryo transfer could be described as a routine experimental tool, but 'The idea that [it] might become a common practice...used in a similar way to artificial insemination...to exploit the genetic potential of the female, is still a long way off'.

There is no doubt that embryo transfer is an extremely useful research tool: it has been used to investigate the degrees to which the mother and fetus or newborn control characteristics such as gestation length, birth weight, fleece qualities and immunoglobin absorption from colostrum; it continues to add to our understanding of the interrelationships between the embryo, uterus and ovary that are essential to establish and maintain pregnancy; and it is an essential component of all embryo manipulation procedures.

The striking development since 1972, though, is that embryo transfer in cattle has indeed 'become a common practice'. In that time, in North America alone it has grown from nothing to an industry that transferred over 143000 embryos (about 20 per cent of them frozen–thawed) to produce about 85000 calves in 1983. It is also a well-established commercial procedure in Europe and Australasia on a somewhat smaller scale, and has been used for international shipment of cattle, sheep, goats, pigs and horses. It has been used in the production of calves, lambs, kids and foals from frozen embryos; in the production of identical twin calves, lambs, pigs and foals and even identical quintuplet lambs; to produce calves and lambs of known sex; to produce calves, lambs and pigs from oocytes fertilized *in vitro*; to produce chimaeric animals comprising cells from sheep and goats, and sheep and cattle; to produce interspecific pregnancies (lambs have been born to goats, kids to ewes, donkeys and even a zebra to mares, and a gaur calf to a cow); and most recently, in the production of transgenic livestock (for example, pigs and sheep into which foreign genes were injected at the pronuclear stage). It has been used to treat some kinds of infertility and for certain disease control measures (e.g. the introduction of new blood lines into specific-pathogen-free pig herds). The AI industry has made direct use of embryo transfer to test its bulls for Mendelian recessives, and many AI stud bulls are being derived from contract transfers. In short, embryo transfer has come a long way in 14 years and is continuing to develop extremely rapidly. Consequently, anyone interested in the future of animal production should know something of its potential and its limitations.

Commercial use of the technique in cattle by specialized embryo transfer units began in the early 1970s and was largely spurred on by the profitability of multiplying 'exotic' European breeds of cattle that had been imported into North America and Australasia in limited numbers under quarantine regulations. Commercial incentives have contributed to rapid technical advances in transfer procedures. Successful embryo transfer involves many steps, all of which need close attention to detail (Fig. 1.10).

Techniques, especially those of transcervical collection and transfer in cattle, demand considerable practice.

Cattle embryos are now collected from donors through the cervix ('non-surgically'), usually as late morulae or blastocysts 6–8 days after insemination. Donors are generally induced to superovulate by treatment

Fig. 1.10. Synopsis of bovine embryo transfer procedures. (From G. E. Seidel, Jr. Superovulation and embryo transfer in cattle. *Science*, **211**, 351–8, fig. 2 (1981).)

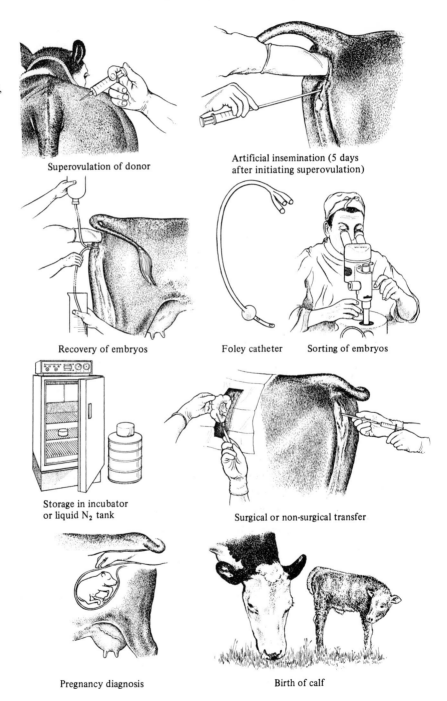

Superovulation of donor

Artificial insemination (5 days after initiating superovulation)

Recovery of embryos

Foley catheter   Sorting of embryos

Storage in incubator or liquid N$_2$ tank

Surgical or non-surgical transfer

Pregnancy diagnosis   Birth of calf

during the luteal phase of the cycle with gonadotrophin (eCG or FSH–LH) followed by prostaglandin $F_{2\alpha}$ (or an analogue) to cause luteolysis and oestrus. Individual variability in response to superovulatory treatments is still a major problem; although an average yield of five transferable embryos can be expected from each responding donor, it may range from zero to more than 20 embryos in different animals. The proportion of treated animals that respond to gonadotrophins varies between 60 and 90 per cent. Hormonal treatment of high-yielding lactating dairy cows can result in milk loss, leading some units to concentrate on repeated collections of single embryos during successive, untreated cycles.

Transcervical collection devices are normally based on a fine rubber Foley-type catheter with a balloon cuff that is introduced under aseptic conditions through the cervix into the base of a uterine horn. Sterile flushing medium (most often an enriched, phosphate-buffered saline) is run into the horn by gravity or from a syringe until adequate distension can be felt *per rectum*. Infusion is then stopped and the medium collected back through the same catheter into suitable vessels. This procedure is repeated several times on each horn.

To retrieve the embryos from the flush, its volume is first reduced to a convenient amount that can be thoroughly searched in Petri dishes under a dissecting microscope. The volume can be reduced by siphoning off most of the medium after a period of sedimentation because embryos gravitate to the bottom of the collection vessel. However, it is now more usual to pass the flush through a plankton filter that retains the embryos which are then rinsed off into the Petri dishes. Once located under the microscope, the embryos are transferred with a Pasteur pipette to a smaller volume of fresh medium for closer examination. Those assessed as morphologically normal (Fig. 1.11) are held in medium at room temperature or at 37 °C until they are transferred to recipients, or prepared for more specialized treatment such as sexing or freezing.

Transfer is still sometimes performed surgically through a flank incision in standing animals under regional anaesthesia. The uterine horn is punctured with a blunted probe to admit a Pasteur pipette, from which the embryo is expelled into the uterine lumen with a small volume of medium. However, non-surgical transfer with an AI gun (for example, the Cassou or the Hanover models) passed through the cervix is now far more widely used because of its simplicity. Some commercial groups consider that it can, with much practice, give results as good as those obtained surgically, but most agree that pregnancy rates are slightly lower and more variable with non-surgical methods. With either technique it is also possible to transfer an additional embryo to mated recipients to induce twinning or improve the chances of single pregnancy.

Recipients' oestrous cycles must be closely synchronous with the donors' for pregnancy to ensue. If fresh embryos are being transferred, this necessitates having a large herd of cycling cattle to provide sufficient

Fig. 1.11. Cattle embryos at various transferable stages of development. In practice, only those illustrated in photos (*c*)–(*f*) would be used; these depict embryos recovered on days 6–8 (oestrus = day 0). (*a*) Four-cell egg, day 3. (*b*) Sixteen-cell egg, day 5. (*c*) Morula, day 6. Cells have compacted and lost their visible individual outlines. (*d*) Early blastocyst, day 7. Note the prominent inner cell mass. (*e*) Blastocyst, fully expanded within the zona pellucida, day 10. The inner cell mass is still obvious. (*f*) 'Hatching' blastocyst, day 10. Note the cells extruding through the zona pellucida. (*g*) 'Hatched' blastocyst, day 12, with typical wrinkled appearance. (*h*) Elongating blastodermic vesicle, day 14, with a prominent embryonic disc. Part of a much larger embryo from the same donor can be seen at the bottom of the photograph. (Photographs from the Animal Diseases Research Institute, Agriculture Canada, Ottawa.)

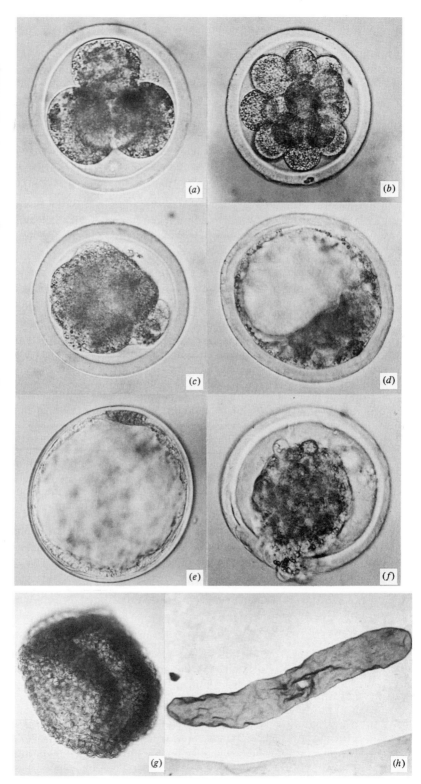

recipients on any given day, an expense that constitutes a major part of the cost of embryo transfer. The herd size can be reduced by using artificial synchronization of oestrous cycles in some circumstances, or by thawing frozen embryos as recipients become available. Following direct transfer of single embryos to the uterine horn ipsilateral to the corpus luteum, 50–70 per cent of recipients become pregnant; with twin transfers (one embryo to each horn) pregnancy rates range between 70–90 per cent, with 40–60 per cent embryo survival.

Techniques and results in sheep and goats are basically similar to those described for cattle, except that surgical methods have to be used for embryo collection, and surgery or laparoscopy for transfer. Indeed, sheep have been, and continue to be, useful models in which to develop procedures and applications. In Australia, proliferation of Angora goats by the use of feral scrub goat recipients for Angora embryos has become of commercial interest.

Embryo transfers in pigs also have to be made surgically. Within 7 days of ovulation, embryo survival rates are similar to those occurring naturally. Pigs can be readily superovulated, but methods of synchronization are more complicated than in cattle. Fortunately, synchrony is less of an obstacle to successful embryo transfer in pigs than in cattle, for Chris Polge and his colleagues at Cambridge have demonstrated that, as long as the recipient ovulates after the donor, 2 or 3 days' asynchrony does not seriously affect results and even a 4-day 'younger' uterus can carry embryos to term in a good proportion of cases. Pig embryos proved extremely refractory to attempts to preserve them by cooling and freezing.

Mares are especially difficult to superovulate and have yet to be made to produce foals from transferred multiple embryos. However, mares are ideal for non-surgical collection and transfer methods because of the straight, easily dilatable cervix. Repeated collection of single embryos is therefore quite feasible. As in pigs, the need for synchrony between donor and recipient in mares is apparently less rigid than in cattle and may be made obsolete if initial results in progesterone-treated ovariectomized or intact recipients live up to their promise. Although practical application of embryo transfer in horses is still limited by the registration requirements of many breed societies, other horse breeders are making increasing use of the procedure.

It remains true that the costs of embryo transfer will probably never be low enough to make it a production tool that can be used to the same extent as artificial insemination. Also, the number of progeny per female will always be about a thousand-fold smaller than for the male in Foote's second equation, Equation 1.2 (page 17). However, rapid and continuing progress has consolidated its position as a unique means of meeting certain production and many research requirements in farm species.

Embryo preservation. Embryos of cattle and other species can be held at room temperature or in an incubator at 37 °C during the day of transfer without evidently reducing their viability. When held in a refrigerator at 10 °C, a reasonable proportion remain viable after about 2 days, and an occasional one survives transfer after up to 10 days' storage. The oviducts of rabbits and sheep have also proved useful 'biological incubators' for experimental work. However, for practical purposes, deep-freezing in liquid nitrogen at −196 °C is required for embryos to be 'banked' indefinitely, and transported and transferred as required.

Efforts to preserve embryos by freezing began at about the same time as semen freezing, but no young were produced until the early 1970s. A basic difference between freezing semen and freezing embryos is that a vial or straw with one killed embryo cannot produce a pregnancy, whereas a vial or straw with millions of killed spermatozoa can, provided enough living ones remain. Freezing embryos involves carefully controlled cooling rates and the use of 'cryoprotective' agents (such as dimethylsulphoxide, glycerol, propanediol, methanol or ethylene glycol) which allow balanced dehydration of cells so that they are not damaged either by intracellular ice crystals or by adverse effects on cell membranes. Techniques and apparatus for freezing and thawing embryos in ampoules or straws have been improved and simplified to the point where careful use of published recipes on good quality bovine embryos produces pregnancy rates that approach those obtained by transferring fresh embryos. Methods for freezing bovine embryos in straws allow thawing and transfer of the embryo to be done on the farm, very much like AI. Commercial use of cryopreservation is therefore already routine and the proportion of bovine embryos that are frozen is undoubtedly increasing from the 1983 level of 20 per cent.

Embryo sexing. The ideal method of predetermining the sex of farm animals would be to separate living X- and Y-bearing spermatozoa. However, despite frequent claims to the contrary, this ideal has never been convincingly achieved and may, in the words of one of the editors of this series, 'rank amongst Nature's impossibilities'. The only methods of determining sex prenatally have therefore been applied to fetuses sampled by amniocentesis (which is often at too late a stage to be practical) or to embryos during transfer.

Cattle and sheep embryos have been successfully sexed and transferred after hatching from the zona pellucida, when a piece of trophectoderm can be removed for chromosomal analysis without damaging the embryonic disc. Unfortunately, these hatched embryos cannot be frozen, which limits the value of the technique. Attempts to perform chromosome analysis on blastomeres removed with a micropipette inserted through the zona pellucida of younger, freezable embryos have produced conflicting results, and that method cannot yet be considered practical. Calves of known sex have been produced from frozen-thawed half-embryos produced by micromanipulation (see below), so that the other half can be used for chromosome analysis (Fig. 1.12). Although reliable, this technique is unsuitable for general application because it is too complex and because about one-third of embryos cannot be analysed. An alternative approach involves a male-specific antigen (which may or may not be H-Y antigen, described in Book 2, Chapter 3, Second Edition) which can be detected immunologically on the surface of entire embryos. This looks promising, for with it Gary Anderson's group in Davis, California, have accurately identified the sex of over 80 per cent of mouse, cow, pig and sheep embryos. However, at present, the test remains too subjective for the transition from laboratory to field. A further possibility has been demonstrated in mice: assay of the X-linked enzyme glucose-6-phosphate dehydrogenase by a simple *in vivo* dye test has been used successfully to sex embryos during the preimplantation days before X-inactivation, when there is a twofold difference in the levels of the enzyme activity between the female and male embryos (Book 2, Chapter 1, Second Edition). The same principle holds for farm animals, but has yet to be proved a suitable basis for practical sexing. Whatever sex-determining measures do prove best in the near future, embryo transfer seems likely to be involved.

Embryo micromanipulation. Advanced micromanipulation of mammalian embryos (see Book 2, Chapter 6, Second Edition) is no longer confined to laboratory species. The work of Steen Willadsen and his colleagues in Cambridge can surely be regarded as 'trend-setting' (Fig. 1.13). The

Fig. 1.12. The use of embryo splitting to produce calves of known sex. (*a*) and (*c*) show the two halves obtained from bisection of a single blastocyst and then inserted into opened zonas (cf. Fig. 1.14). Half-embryo (*a*) was frozen, thawed and transferred to produce a calf (*b*). Meanwhile, half-embryo (*c*) was cultured in the presence of Colcemid and used to obtain a cytological preparation (*d*), in which the metaphases are marked by arrows. Chromosome analysis of these metaphases showed that the embryo was male. This bull-calf (*b*) was the first to be born from a frozen, sexed embryo. (From L. Picard, W. A. King and K. J. Betteridge. *Veterinary Record*, **117**, 603–8 (1985).)

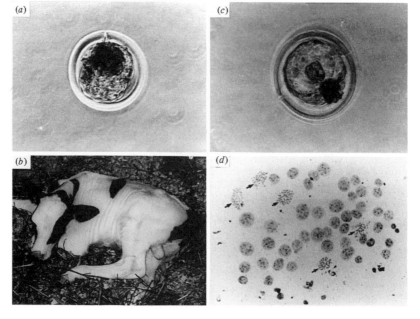

Fig. 1.13. Four stages in the production of identical twin foals by micromanipulation of embryos (the work of S. M. Willadsen, R. L. Pashen and W. R. Allen. See W. R. Allen and R. L. Pashen. *J. Reprod. Fert.* **71**, 607–13 (1984).) (*a*) A two-cell egg is recovered surgically from the donor mare and the zona pellucida is opened by micromanipulation to release and separate the blastomeres. (*b*) The separate blastomeres are injected into emptied zonas derived from porcine follicular oocytes. The newly constituted embryos are embedded in agar cylinders to seal the opening in the zona. (*c*) The embedded embryos are surgically inserted into the ligated oviduct of a sheep and allowed to develop into blastocysts during 4 days' incubation. (*d*) After recovery from the sheep, the blastocysts are transferred surgically into the uteri of two mares that ovulated at about the same time as the donor. Identical twin foals are produced by these recipient mares after 'normal' gestations.

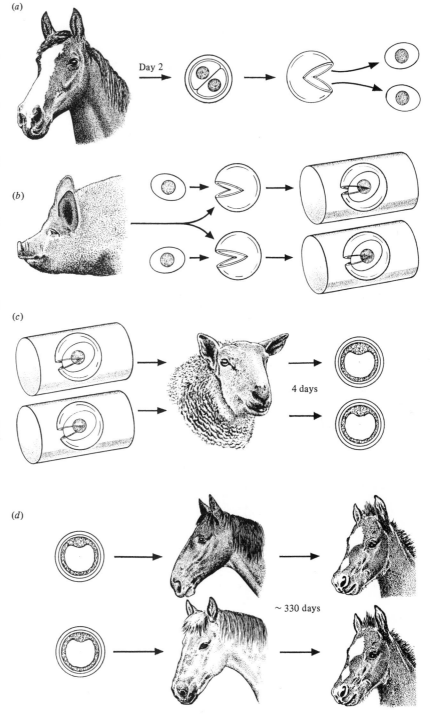

separation of the blastomeres of pre-compaction sheep and cow embryos has led to the birth of identical quadruplet lambs and triplet calves.

Equally remarkable is the speed with which 'embryo splitting' passed from the developmental laboratory to the field, once it was shown by Willadsen and others (notably Jean-Paul Ozil and his colleagues in France, Tim Williams and his colleagues in Colorado, and Bob Godke's group in Louisiana) to be practicable on bovine post-compaction morulae and blastocysts recovered from the uterus. There are now several versions of the bisection technique and they have been used successfully in cattle, sheep, goats, pigs and horses. The bisection is accomplished under the microscope with the morula or blastocyst, within its zona pellucida, held by suction as described in Book 2, Chapter 6, Second Edition. A piece of razor blade or a fine glass needle is used to divide the embryo into two equal halves (through the inner cell mass in the case of a blastocyst), either within the zona pellucida or after removal of the zona from around the embryo (Fig. 1.14). Half-embryos rapidly re-compact, seal, and form (or reform) a blastocoele. They can be transferred to recipients either directly

Fig. 1.14. Diagrammatic representation of the 'embryo splitting' procedure that is now widely used in commercial embryo transfer practice.

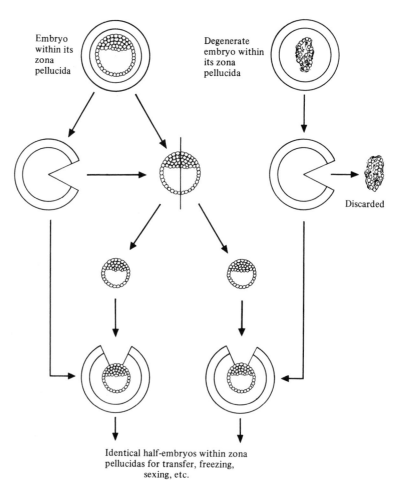

Embryo within its zona pellucida

Degenerate embryo within its zona pellucida

Discarded

Identical half-embryos within zona pellucidas for transfer, freezing, sexing, etc.

or after insertion into an emptied zona obtained from the original embryo or from an unfertilized egg. Further sub-division of post-compaction embryos does not seem useful at present because no more than twins have been obtained from day-6 embryos split into four. Several commercial embryo transfer companies now offer embryo bisection as a routine service and, because the number of calves produced following transfer of freshly split embryos can exceed the original number of embryos used (pregnancy rates of 100–150 per cent are reported), some companies split nearly all recovered embryos before transfer. Identical twins survive from 25–50 per cent of split embryos. There is no doubt that embryo splitting has also increased the value of embryo transfer to genetic improvement programmes (see Nicholas and Smith, 1983, in the Suggested further reading list) and provided a valuable source of animals for experimental studies requiring identical treatment and control groups. Treating one half-embryo as a 'biopsy' has been used for sex diagnosis, as described above, and could also be useful for monitoring embryos for chromosomal abnormalities and the presence of infectious agents, both factors of importance when international shipment of embryos is being considered. To realize the full potential of embryo splitting, it will be important to improve present techniques for freezing divided embryos.

Another aspect of micromanipulation that has made the transition from laboratory to farm as a result of Steen Willadsen's and Carole Fehilly's work is the production of chimaeras (see Book 2, Chapter 6, Second Edition). The somewhat bizarre animals that result from the mixing of sheep and goat or of sheep and cow blastomeres (Fig. 1.15) have occasioned uninformed criticism from champions of 'animal rights', so the

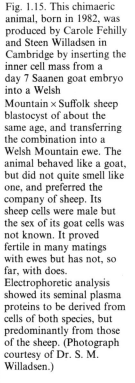

Fig. 1.15. This chimaeric animal, born in 1982, was produced by Carole Fehilly and Steen Willadsen in Cambridge by inserting the inner cell mass from a day 7 Saanen goat embryo into a Welsh Mountain × Suffolk sheep blastocyst of about the same age, and transferring the combination into a Welsh Mountain ewe. The animal behaved like a goat, but did not quite smell like one, and preferred the company of sheep. Its sheep cells were male but the sex of its goat cells was not known. It proved fertile in many matings with ewes but has not, so far, with does. Electrophoretic analysis showed its seminal plasma proteins to be derived from cells of both species, but predominantly from those of the sheep. (Photograph courtesy of Dr. S. M. Willadsen.)

potential importance of the work to animal production should be understood. When chimaeras are made by combining one blastomere from a four-cell embryo and one blastomere from an eight-cell embryo, the older ($\frac{1}{8}$) cell tends to form the embryo proper while the younger ($\frac{1}{4}$) cell forms the trophoblast (Fig. 1.16). Thus, in theory, eight identical individuals could be produced; in practice to date, the Cambridge workers have produced identical quintuplet lambs. When blastomeres from different species are combined in this way, it can be seen that it is possible to arrange for an inner cell mass of one species to develop within the trophoblast of another. Because it is the trophoblast that determines the acceptability of the conceptus to the mother, this technique allows young to be produced from interspecies transfers that would otherwise fail (between sheep and goats, for example). This could have important implications for the conservation of rare species and thence, in the long term, for agriculture. Not all interspecies transfers require such elaborate procedures; the zebra (Fig. 1.17) and donkey foals and gaur calf referred to earlier were produced by direct transfers. In the shorter term, it is worth remembering that we know nothing of the production characteristics of interbreed, intersex, or interspecies chimaeric livestock.

As in laboratory species (discussed in Book 2, Chapter 6, Second Edition), the potential value of mixing individuals in farm species is now being investigated at the sub-cellular level, by nuclear transplantation. Lambs have recently been produced in this way, but work with amphibians suggests that the nuclei will have to be embryonic if they are to be capable of controlling embryonic differentiation. The more extravagant claims concerning 'cloning' of valuable adult animals are therefore quite unrealistic at present; it is only unproven embryos that can be duplicated by any existing, or immediately forseeable, technique.

A concomitant research thrust in this rapidly advancing field is at the sub-nuclear level, i.e. the creation of 'transgenic' livestock by microinjection of genes themselves into a pronucleus soon after fertilization and before the first cleavage of the zygote. Some measure of the research effort being invested in this mammalian aspect of the new 'biotechnology' is given by the participation of over 300 delegates from 11 countries at a four-day conference entitled 'Genetic Engineering of Animals: An Agricultural

Fig. 1.16. Manufacture of 'three-eighth' sheep embryos. One cell from an eight-cell and one cell from a four-cell embryo are placed together in a surrogate zona pellucida. In most cases, inner cell masses arise exclusively from the blastomere originating from the eight-cell embryo. (From S. M. Willadsen and C. B. Fehilly, cited by G. E. Seidel, Jr. *Biol. Reprod.* **28**, 36–49 (1983).)

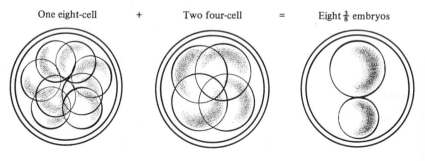

One eight-cell   +   Two four-cell   =   Eight $\frac{3}{8}$ embryos

Perspective' held in Davis, California, in September 1985. The interest stems from the successful introduction of foreign genes into mouse eggs in the early 1980s: first, genes for thymidine kinase and human β-globin were transferred by Tom Wagner and his colleagues in Ohio, then both human and rat growth hormone genes were transferred by Richard Palmiter, Ralph Brinster and their colleagues in Philadelphia. In the latter study, the promoter region of the mouse gene for metallothionein was fused to the structural gene coding for growth hormone to produce fusion genes for injection into the male pronucleus where, because of DNA repair

Fig. 1.17. A Grant's zebra foal with its quarter horse surrogate mother, Kelly, to which it had been transferred as a 10-day-old blastocyst by Drs William R. Foster and Scott D. Bennett. This male foal was born at the Louisville Zoo in Kentucky on 17 May 1984, 356 days after transfer. (Photograph by Paul Schuhmann, Copyright © 1984, *The Courier-Journal.* Reprinted with permission and through the co-operation of Mary T. Duane of the Louisville Zoo.)

activity, some were incorporated into the mouse embryo's chromosomes. The small proportion of mice in which the genes were incorporated produced high levels of growth hormone, grew significantly faster and bigger than control littermates, and made more efficient use of food. The foreign genes were 'switched on' by feeding zinc, which induces the expression of the metallothionein gene in normal mice. Some introduced genes were expressed in subsequent generations but there was also considerable infertility in the transgenic individuals. Since then, more moderate improvements in growth rates have been achieved by introduction of genes for growth-hormone-releasing hormone, and improved growth seems to be inherited in a more reproducible fashion. Metallothionein–hGH fusion genes have now been integrated into the genomes of sheep, rabbits and pigs, and these were expressed in rabbits and pigs but not sheep. Prospects for sheep are discussed by Kevin Ward and his colleagues (see Suggested further reading). Factors controlling the incorporation of injected material are still poorly understood and it seems likely that we shall look back on these efforts in a few years' time as having been early hit-and-miss attempts in a field of immense potential value to animal production.

In vitro *fertilization*. The spectacular advances that have been made in human *in vitro* fertilization (IVF) in the last decade (discussed in Chapter 5 of this book) owe a great deal to work in both laboratory and farm species, so it is surprising that progress in producing farm animals themselves by the procedure has been relatively slow. Thus, the first IVF calf (Fig. 1.18) was born in Pennsylvania as recently as 1981, after

Fig. 1.18. This 48 kg bull-calf, Virgil, seen here with his surrogate dam, Penny, was born on 9 June 1981 at the University of Pennsylvania's School of Veterinary Medicine, New Bolton Center. It was the first calf to result from *in vitro* fertilization and embryo transfer. (Fig. 3 in B. G. Brackett, D. Bousquet, M. L. Boice, W. J. Donawick, J. F. Evans and M. A. Dressel. *Biol. Reprod.* **27**, 147–58 (1982).)

years of preliminary work by Ben Brackett and his colleagues, and the first IVF pigs and lambs not until 1983 and 1984, respectively, in Cambridge, where Winston Cheng showed what a difference 2° C could make to the procedure! Even now (1986), calves are arriving only slowly and sporadically. An intermediate incubation of newly fertilized bovine ova in the sheep or rabbit oviduct before transfer to the recipient cow has helped investigators at Laval (Quebec) and in Wisconsin to make IVF results a little more consistent, but further improvement is needed to provide uncleaved zygotes routinely for gene injection, for example.

Diagnosis of pregnancy and non-pregnancy
It is economically and reproductively inefficient to keep female farm mammals if they are not pregnant when they should be: for example, the 17 per cent or so of beef cows that were not pregnant at the end of the breeding season in the disease-free herd documented in Table 1.4. It has been calculated that culling such non-pregnant cows could save winter feeding costs of over \$280 million per year in the USA. Analogous comments apply to other farm species, especially with intensive systems of management, so sorting the pregnant from the non-pregnant is of major importance. For this purpose, it is often as useful to be able to identify non-pregnant animals as to be able to diagnose pregnancy.

Methods of pregnancy diagnosis in farm animals have been tabulated in Book 3, Chapter 7, Second Edition. The choice of method will depend on the exact needs and circumstances of a given production system. For example, in cattle and horses there are often advantages to diagnosis *per rectum* by palpation or ultrasonics, if the skilled manpower and equipment are available: the diagnosis is available immediately, the herd is subjected to close general and veterinary scrutiny, records are kept and, with real-time ultrasound, a photographic proof of pregnancy can be provided (Fig. 1.19). On the other hand, where efficient hormone assay services for pregnancy diagnosis have been set up for the producer, a great deal of reproductive surveillance of a herd or flock becomes possible by intelligent interpretation of results. Conversely, several new hormone assay systems could be used for pregnancy diagnosis if the production system warranted their introduction. For example, in sheep, detection of placental lactogen in blood becomes a reliable indicator of pregnancy about 55 day after mating. This could be useful because it is after the time when most embryonic mortality occurs, so 'false positives' are unlikely. Further, because the test does not depend on an accurate record of breeding dates, it could be done on a whole flock of ewes 55 days after removing the rams. In sows with accurately known breeding dates, non-pregnancy can be diagnosed 13–14 days after insemination when a metabolite of prostaglandin $F_{2\alpha}$ appears in the blood if luteolysis is to occur, but not if pregnancy has been established. In cows, an immunological assay for 'protein B' (one of an array of proteins apparently produced by binucleate cells in the trophoblast) is under commercial development for early pregnancy diagnosis, and other groups may follow suit with other trophoblastic proteins. Trophoblastic and placental hormones are of especial interest in the diagnosis of pregnancy because they provide direct evidence of the presence of a conceptus. For research purposes, this is of value in studies on embryonic or fetal death; from a population standpoint they may eventually be useful for differentiating between twin and single pregnancies. Thus, in cattle, even though placental oestrogens only give this indication after 220 days' gestation, such an assay could allow the feeding of the dam to be properly adjusted and so improve birthweights and early growth. Even confirmation of pregnancy near term can have its place; udder palpation is used in programmed sheep breeding to decide which ewes could respond to treatment for the induction of parturition.

Control of parturition and reduction of perinatal loss

One of the reasons for the advances in our understanding of the physiology and endocrinology of parturition in farm animals has been the use of sheep

Fig. 1.19. Ultrasonograms showing the development of the equine conceptus during the first 7 weeks of pregnancy.
(*a*), (*b*) and (*c*): Normal development of the same spherical conceptus within the uterus of a Thoroughbred mare as seen at 9.5, 11.5 and 14.5 days (±0.5 day) after ovulation. The diameter of the conceptus on these respective days was 2, 7 and 20 mm (see also Book 2, Second Edition, Fig. 2.18). This particular embryo had been transferred as a 0.49 mm blastocyst on day 7.5.
(*d*) and (*e*): the ultrasonogram and its diagrammatic interpretation in another mare at 47.5 days' gestation, when the fetus and its movements are clearly visible. This conceptus is of additional interest because it had been removed from the uterus at day 10.5, examined, and then reinserted non-surgically to continue its development.
   Ultrasonography not only provides a graphic record of the presence of an embryo but has also (notably through the work of O. J. Ginther) taught us a great deal about embryo growth, movement, orientation and fixation within the uterus during early pregnancy and in cases of early embryonic death.
(Photographs courtesy of Jean Sirois, CRRA, Faculté de Médecine Vétérinaire, Université de Montréal.)

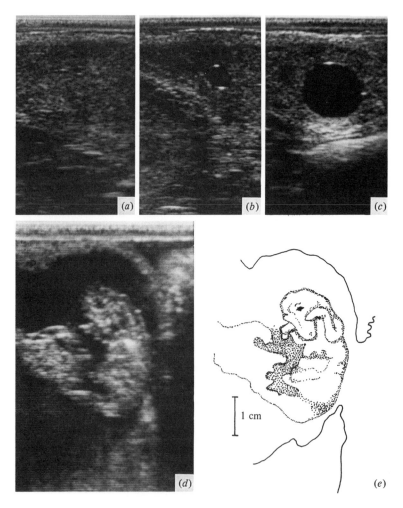

1 cm

and goats as models for investigation of human parturition. Human medicine has made use of the results of this research, with consequent reductions in the incidence of perinatal mortality. Paradoxically, the levels of perinatal mortality in farm animals remain high; so high that it is surely time to take steps to reduce them. The scale of the problem in relation to other reproductive losses in beef cattle is shown in Table 1.4. Other surveys indicate that pre-weaning mortality for calves ranges between 8 and 20 per cent, for lambs around 20 per cent, piglets 18–26 per cent and foals about 5 per cent. Most of the losses are not associated with infectious disease, but occur close to the time of birth, when the fetus has to pass through the physiological changes associated with adaptation from an intra- to an extra-uterine existence. These adaptive changes are beyond the purview of this chapter but they are influenced directly by the endocrine changes involved with parturition, which is very much part of reproductive biology (see Book 2, Chapter 4, Second Edition).

Research into fetal physiology should eventually provide the information we need if neonatal mortality is to be reduced. Such work has already led to the development of methods for inducing parturition in all domestic species. The practical value of this is that births can now be timed to happen during periods when manpower is available to deal with any problems that occur, and hence to prevent losses due to prolonged labour or poor mothering. Generally, induction of parturition should only be used close to normal term, otherwise the newborn will be too immature to adapt to its new life and the mother's mammary glands too ill-prepared to feed it. Prematurity is tolerated better by cattle than by other species, and this can be useful where a small calf is desirable for obstetric reasons. Used in conjunction with synchronized breeding, induced parturition can help 'batch farrowing' and 'batch lambing', which can be useful in intensive production systems, enabling entire maternity buildings to be emptied and disinfected at one time.

The hormonal cascade, starting in the fetus, that brings about delivery has been described elsewhere in this series (Book 2, Chapter 4, Second Edition). Various hormones from that cascade have been used for parturition induction in the different species. Glucocorticoids are effective within 2 weeks of term in cattle, 5 days before term in sheep and 3 days before term in pigs. Oestrogens can be used near term in sheep and goats. They sometimes work in cattle, but not in horses or pigs. Prostaglandin $F_{2\alpha}$ or its analogues are effective in all species, as exemplified for the pig in Fig. 1.20. Induction of parturition in mares is usually accomplished with oxytocin infusions, though relatively high doses of glucocorticoids may also be effective.

Reduction of embryonic and fetal losses
Unfortunately, there is little progress to report in the prevention of embryonic mortality. Indeed, it is often regarded as a normal, in-built

character of the species concerned (see Book 4, Chapter 2, Second Edition). However, increasing knowledge of how interactions between the embryo and the maternal organism evolve during the establishment of pregnancy suggests that artificial control of this relationship may eventually be useful for reducing embryonic mortality in selected farm animals, at least in the special circumstances following embryo transfer. Who could have forecast, for example, that ovariectomized mares would make good recipients, so long as they were treated with progesterone? It is also challenging to note that Eric Bradford has been able to reduce embryonic mortality in laboratory mice through genetic selection. This reduction is evidently effected through improvement of the uterine environment, rather than by changing the quality of the embryos (Book 4, Chapter 3, Second Edition).

The extent and economic importance of embryonic mortality, first established in slaughtered animals, has been confirmed through the use of early tests for pregnancy (particularly milk progesterone assays in cattle and ultrasonics in mares) after insemination or embryo transfer. Its pre-eminence as a reproductive problem in cattle is seen in Table 1.5 (p. 7). In sheep, approximately 10–30 per cent of embryos die, the higher death rates occurring early in the breeding season; in pigs, the figure is about 33 per cent, and, in mares, it is about 15 per cent after the second week of pregnancy. Transfer work has shown embryonic mortality to be accentuated by storage and embryo manipulation, and it is worth noting that transfer procedures producing pregnancy rates of 60 per cent are also inducing embryonic mortality rates of 40 per cent. Losses are not confined

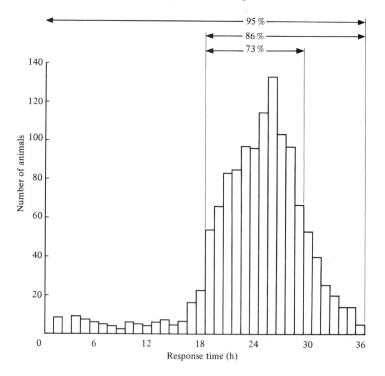

Fig. 1.20. Distribution of farrowings in 1355 sows treated with cloprostenol, an analogue of prostaglandin $F_{2\alpha}$, 1, 2, 3, or 4 days before the expected farrowing date. (From M. Cooper. Prostaglandins in veterinary practice. *In Practice*, **3**, 30–4 (1981).)

to any one critical stage, and those occurring before day 16 in cattle and day 12 in sheep go unrecognized, indistinguishable from fertilization failure, because the dam returns to oestrus at the normal time. Recent cytogenetic studies suggest that chromosome abormalities associated with fertilization and early cleavage may be more common and more important in farm animals than was previously thought.

Programmed breeding

Figure 1.21 shows how the foregoing components of reproductive control systems can be integrated for food production. The French have been pioneers in intensive sheep breeding, and their programmes, or modifications of them, are proving feasible in several other countries. The economics of these systems will be affected by factors such as climate, which dictates whether permanent housing is necessary. What is important in the present context is to realize that intensive food production can be achieved even with seasonally breeding animals. It is therefore possible that such systems will serve as a model for the development of analogous programmes in other species.

In summary, farm animals remain indispensable to food production for man and we have seen some of the ways in which reproductive biology can contribute to increasing their productivity. There is much to be done in making better use of existing knowledge; dissemination of currently available information has been woefully slow, even in industrialized countries, and it is humiliating to realize that something as elementary as the detection of oestrus in dairy cattle is one of the greatest impediments to increasing the efficiency of the industry. The development of artificial insemination and the deep-freezing of semen has had an enormous impact in cattle breeding, although it has been of limited significance in sheep, pigs

Fig. 1.21. A moderately intensive flock management scheme for lamb production that produces three lambs per female per year. New batches of ewes are introduced into the scheme at 49-day intervals; interval between lambing and next AI = approximately 52 days. (From J. Thimonier, Y. Cognie, C. Cornu, J. Schneberger and G. Vernusse. Intensive lamb production. *Ann. Biol. Anim., Biochim. Biophys.* **15**, 365–7 (1975).)

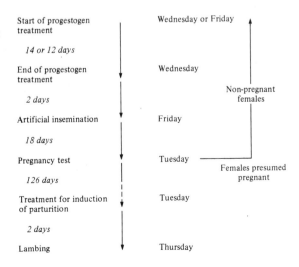

and horses. Embryo transfer and associated techniques have already made important contributions of a more specialized nature in pedigree herds, are invaluable research tools, and will be essential components of the future exploitation of *in vitro* fertilization and 'bioengineering' by gene injection. It is interesting to see how, so far, research in all these areas has pre-dated human clinical application of the techniques, often by several decades. There can be little doubt that it has been the animal example that has led ultimately to the medical advance in the treatment of human infertility. However, there are also important comparisons to be drawn between those of us who seek to increase fertility in animals, and those who attempt to reduce it for human population control. Thus, a phenomenon like lactational anovulation can be a nuisance to the pig, dairy or beef farmer, a blessing to the family planner, and a mechanism of absorbing interest to the basic reproductive biologist.

Suggested further reading

An historical look at embryo transfer. K. J. Betteridge. *Journal of Reproduction and Fertility*, **62**, 1–13 (1981).

Advances in the physiology of growth, reproduction and lactation. W. Hansel. *Cornell Veterinarian*, **75**, 56–76 (1985).

Cloning mammals: current reality and future prospects. C. L. Markert. *Theriogenology*, **21**, 60–7 (1984).

Increased rates of genetic change in dairy cows by embryo transfer and splitting. F. W. Nicholas and C. Smith. *Animal Production*, **36**, 341–53 (1983).

Perinatal mortality: some problems of adaptation at birth. G. C. B. Randall. *Advances in Veterinary Science and Comparative Medicine*, **22**, 53–81 (1978).

Programmed year-round sheep breeding. T. J. Robinson. *Australian Journal of Experimental Agriculture and Animal Husbandry*, **20**, 667–73 (1980).

Superovulation and embryo transfer in cattle. G. E. Seidel, Jr. *Science*, **211**, 351–8 (1981).

Mammalian oocytes and preimplantation embryos as methodological components. G. E. Seidel, Jr. *Biology of Reproduction*, **28**, 36–49 (1983).

Equine embryo transfer. Symposium, ed. W. R. Allen and D. F. Antczac. *Equine Veterinary Journal*, Supplement **3** (1985).

The survival of transferred embryos. Symposium, International Embryo Transfer Society. *Theriogenology*, **23**, 45–174 (1985).

Melatonin: a co-ordinating signal for mammalian reproduction? L. Tamarkin, C. J. Baird and O. F. X. Almeida. *Science*, **227**, 714–20 (1985).

The possibility of transgenic livestock. T. E. Wagner, F. A. Murray, B. Minhas and D. C. Kraemer. *Theriogenology*, **21**, 29–44 (1984).

Nuclear transplantation in sheep embryos. S. M. Willadsen. *Nature*, **320**, 63–5 (1986).

Animal production and the world food situation. J. H. G. Holmes. In *World Animal Science. A.1: Domestication, Conservation and Use of Animal Resources*, pp. 255–84. Ed. L. Peel and D. E. Tribe. Elsevier; Amsterdam (1983).

The role of embryo gene transfer in sheep breeding programmes. K. A. Ward,

J. D. Murray, C. D. Nancarrow, M. P. Boland and R. Sutton. In
Reproduction in Sheep, pp. 279–85. Ed. D. R. Lindsay and D. T. Pearce.
Australian Academy of Science and Australian Wool Corporation (1984).
Current use of frozen boar semen – future need of frozen boar semen.
H. C. B. Reed. *Proceedings of the First International Conference on
Deep-freezing of Boar Semen. 25–27 August 1985.* Uppsala; Sweden (in press).
Beltsville Symposia in Agricultural Research. (3) Animal Reproduction. Ed.
H. W. Hawk. Allenheld, Osmun & Co.; Montclair (1979).
*Current Therapy in Theriogenology: Diagnosis, Treatment and Prevention of
Reproductive Diseases in Animals.* Ed. D. A. Morrow. W. B. Saunders;
Philadelphia (1980). (Second edition in press, 1985.)
Maternal Recognition of Pregnancy. Ciba Foundation Symposium 64 (new
series), ed. J. Whelan. Exerpta Medica; Amsterdam (1979).
New Technologies in Animal Breeding. Ed. B. G. Brackett, G. E. Seidel, Jr. and
S. M. Seidel. Academic Press; Amsterdam (1981).
Control of Pig Production. Ed. D. J. A. Cole and G. R. Foxcroft. Butterworth
Scientific; London (1982).
The Male in Farm Animal Reproduction. Ed. M. Courot. For the Commission
of the European Communities. Martinus Nijhoff; Boston (1984).
Genetic Engineering of Animals: An Agricultural Perspective. Ed. J. W. Evans
and A. Hollaender. Plenum Press; New York (1986).
Reproduction in Farm Animals (4th Ed). Ed. E. S. E. Hafez. Lea Febiger;
Philadelphia (1980).
Reproductive Biology of the Mare: Basic and Applied Aspects. O. J. Ginther.
Ginther; Cross Plains, Wisconsin (1979).
The Physiology and Technology of Reproduction in Female Domestic Animals.
R. H. F. Hunter. Academic Press; New York (1980).
*World Animal Science A.1: Domestication, Conservation and Use of Animal
Resources.* Ed. L. Peel and D. E. Tribe. Elsevier; Amsterdam (1983).
World Animal Science. A2: Development of Animal Production Systems. Ed.
B. Nestel. Elsevier; Amsterdam (1984).

2

Today's and tomorrow's contraceptives

R. V. SHORT

I was amazed several years ago to hear Sir Peter Scott of the World Wildlife Fund say, despairingly, that perhaps we would ultimately do more to save endangered animals around the world if the Fund was to devote all its financial resources in future to the purchase and distribution of condoms. I was equally stunned by the thought that the population problem could be solved at a stroke by making cigarettes free, and available from the earliest possible age, to everybody throughout the world. The resultant increases in mortality from low birthweight infants, lung cancer, heart attacks and strokes would not only reduce the population, but would also result in enormous savings on pensions and health insurance in developed countries, as well as redressing their top-heavy age structures, so that they could easily afford to finance such a scheme. Perhaps the cigarettes could be laced with marijuana to make them more pleasurable to smoke, and hence more akin to the 'soma' proposed by Aldous Huxley in his *Brave New World*.

More orthodox attempts to contain the world's population have so far met with varying success; witness the graph showing the growth of the world's population that appeared in the First Edition of this book in 1972, when the figure had reached 3500 million, whereas only 12 years later the figure for 1984 of 4700 million is almost off the page (see Fig. 2.1). United Nations projections of world population growth up until the year 2100, broken down by region, are shown in Fig. 2.2. The numbers have to be plotted logarithmically to make them fit on the page, and it is at once evident that most of the present and future growth will occur in South Asia, Africa and Latin America; the countries of East Asia, notably China, Japan, South Korea and Taiwan, will hopefully have stabilized their numbers by about the year 2025. When added together, we arrive at a total world population of about 10000 million early in the twenty-first century.

How can some of those poverty-stricken, resource-depleted and under-developed countries of Asia, Africa and Latin America sustain a further two- to five-fold increase in their numbers, almost in the space of one man's lifetime, when they are already bulging at the seams? The higher the birth rate, the less chance a poor country has of achieving significant economic progress. If motivation for family planning is dependent on an individual's standard of living, how can we break the vicious cycle of poverty and overpopulation which feed on one another?

A contrary view was expressed by the government of the United States at the recent United Nations-sponsored International Conference on Population in Mexico City in August 1984, where their delegate put forward the bizarre concept that the solution to the world's population problem does not necessarily lie in the provision of family planning services, but rather in the development of free enterprise. In support of his case, he cited the examples of Hong Kong and South Korea, where increases in population have been accompanied by a phase of rapid economic growth. This myopic view was roundly criticized by many delegates from developing countries, who are already far too poor to embark on a South Korean-style economic recovery. Such an attitude is also completely counter to the philosophy of the World Bank, whose President, A. W. Clausen, recently stated:

Fig. 2.1. A page from the 1972 edition of this textbook, when the world's population stood at about 3500 million, updated to show the 1984 figure of 4700 million. By mid 1986, we had already passed the 5000 million mark.

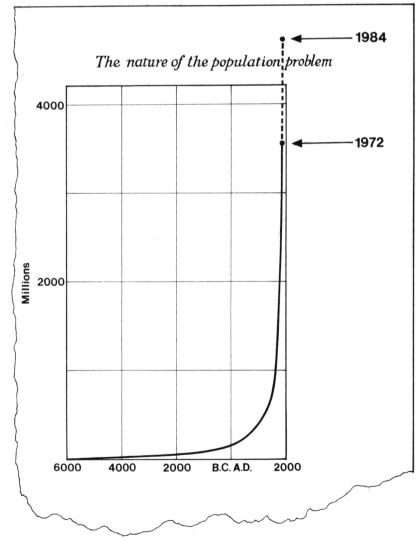

Economic and social progress helps slow population growth; but, at the same time, rapid population growth hampers economic development. It is therefore *imperative* that governments act simultaneously on both fronts. For the poorest countries, development may not be possible at all unless slower population growth can be achieved soon.

Of the three great powers, Russia, China and the United States, it is only China that has shown a positive interest in family planning. The negative attitude of the United States is a major setback, since they are by far the major contributor of funds to global family planning programmes. The recent decision of the United States to withdraw all family planning funds from private organizations that perform or actively promote abortion anywhere in the world is a further blow, which has crippled the activities of the International Planned Parenthood Federation, one of the first casualties. This hypocritical act will do most damage to those developing countries most in need of assistance. No country has been able to control

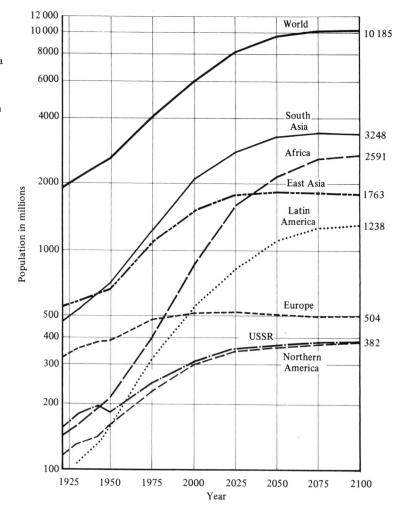

Fig. 2.2. World population growth from 1920 to 2100, broken down by regions, and based on United Nations' estimates. Oceania (Australia, New Zealand and associated islands) is off the bottom of the graph, since the population is projected to reach only 42 million by 2100. Note the logarithmic plot, which obscures the true magnitude of the increases. (From P. Demeny. 1984 and after: can world population forecasts come true? *People*, **11**, 5–8 (1984).)

its rate of population growth without some recourse to abortion (Fig. 2.3, and see also Chapter 3), and the incidence of abortion will only decline with economic development, education, and the provision of adequate contraceptive services. Even in the United States in 1981, there were over 1.25 million legal abortions, representing just over a quarter of all recorded pregnancies (Fig. 2.4); China has an identical proportion of pregnancies ending in abortion, although it is criticized by the United States for its 'liberal' abortion policies. America, why beholdest thou the mote that is in thy brother's eye, but considerest not the beam that is in thine own eye?

To summarize the situation, the world's population will inevitably continue to increase for about another 100 years; our children can therefore expect to inherit a world where there will be twice as many people as there are today. This increase will occur almost exclusively in the developing countries of Asia, Africa and South America. Although it is to be hoped that their birth rates will continue to maintain their steady

Fig. 2.3. No country has been able to control its rate of population growth without recourse to abortion, although the incidence of abortion declines with socioeconomic development and the provision of effective forms of contraception. (From M. Requena. In *Fertility and Family Planning: A World Review*, p. 478. Ed. S. J. Behrman, L. Corsa and R. Freedman. University of Michigan Press (1961).)

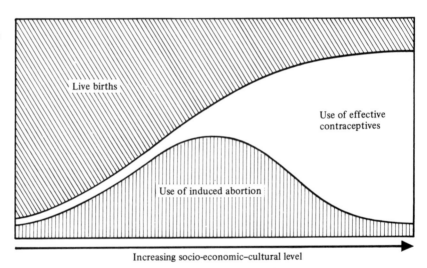

Live births

Use of effective contraceptives

Use of induced abortion

Increasing socio-economic–cultural level

Fig. 2.4 The incidence of abortion in relation to the total number of pregnancies (4.7 million) in the United States in 1981. (From S. J. Segal. Seeking better contraceptives. *Populi*, **11**, 24–30 (1984).)

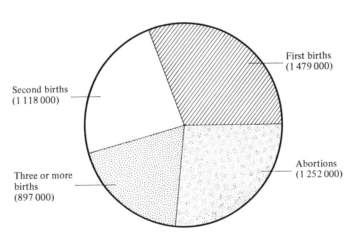

First births (1 479 000)

Second births (1 118 000)

Abortions (1 252 000)

Three or more births (897 000)

decline in response to improved family planning programmes and increased
contraceptive motivation from economic development, the annual increase
in numbers will continue to rise until the end of the century as more and
more people born during the period of high fertility in the 1970s and 1980s
enter their reproductive years (see Fig. 2.5). But that scenario is if all goes
well. If developed countries cut back on their aid for family planning
programmes, as the United States is doing, or if China, representing a
quarter of the population of all developing countries, should falter in her
imaginative attempts to curb her own population growth, then global rates
of growth will exceed projections (see Chapter 7).

It is within this framework that we must begin to consider the topic of
today's and tomorrow's contraceptives. We need hardly discuss the needs
of the developed countries; they have already demonstrated their ability
to regulate their rates of population growth, even though specific groups
such as teenagers still give cause for concern. Developed countries started
to adopt contraception at the end of the nineteenth century, long before
most of the currently used contraceptives were available. But is today's
sophisticated Western contraceptive technology necessarily appropriate
for the developing countries, who are denied the luxury of surplus medical
manpower? (In Britain there is 1 doctor for 760 patients, compared to 1
per 580000 in Nigeria.) More importantly, are Western safety standards
for a contraceptive like Depo-Provera relevant, in view of the pressing
needs of the developing countries for long-acting contraceptives that can
save lives by spacing births?

Fig. 2.5. The crude birth
rate (births per 1000
population) and the annual
increase in population in
developing countries
between 1950 and 2025.
(From P. Demeny. 1984
and after: can world
population forecasts come
true? *People*, **11**, 5–8
(1984).)

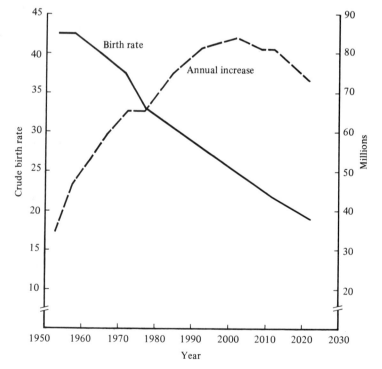

Too often in the past, contraceptive aid has been in the form of piecemeal programmes to promote one or other technique, like oral contraceptive pills, intra-uterine devices, or surgical sterilization. These have been foisted on developing countries by a variety of donor organizations from developed countries that have often been in competition with one another in the rush to recruit acceptors, rather like the unseemly way in which Church missionaries competed for converts in former times. What we need to do now is to formulate overall contraceptive strategies for developing countries; all available methods must be integrated in a way that is culturally most acceptable to the people concerned, and they must satisfy consumer demand. A new spirit of collaboration also needs to be engendered amongst the various funding bodies; fortunately, the World Health Organization, through its Expanded Programme of Research, Development and Research Training in Human Reproduction, has already provided a lead in this area.

If we are to attempt to curb the world's rate of population growth, we need to focus all our attention on Asia, Africa and South America; we need to develop strategies that will have immediate impact, and we cannot afford just to wait for new contraceptive developments that are still only a dream in the mind of the research worker. Time is short, and the waters are rising.

There are three stages in a woman's reproductive life when she may desire contraceptive protection; they are prior to her first pregnancy, between successive pregnancies so as to achieve optimal birth spacing, and after she has completed her desired family. Let us consider each in turn.

Contraception from puberty to first pregnancy

Contraceptive requirements during this stage are strongly dependent on the particular culture. On the one hand there are Islamic and other communities where marriages may be arranged prior to puberty, and the sooner the girl becomes pregnant thereafter, the happier everybody is – except maybe the girl! At the other extreme is the situation such as that in China, where women do not start to marry until their early twenties as a result of intense political and social pressure. Since premarital sex is almost unknown in China, there is no call for contraceptive protection prior to marriage. Somewhere between these two extremes is the Western pattern, where traditionally premarital sex was frowned upon by Church and State, sometimes resulting in the restriction of family planning instruction and contraceptive provision only to those who could provide documentary proof of marriage.

The trend in all developed countries is for an earlier and earlier age at the onset of first intercourse in both sexes, regardless of the age of marriage (see Table 2.1). This same trend is likely to overtake many developing countries, associated with a declining age at puberty as a result of improved nutrition in childhood, and the relaxation of religious, social and cultural taboos against premarital sex. However, the emancipation of women, enabling them to benefit from secondary and even tertiary education, and their growing importance in the urban work force, all encourage a later onset of childbearing. Hence, there will be a growing need in developing countries for access to reliable forms of contraception during this phase of a woman's reproductive life.

What contraceptives are best suited to this group? Clearly, no contraceptive that compromises subsequent fertility should be contemplated; male and female sterilization are automatically ruled out, and long-acting steroidal implants and injections whose effects are only slowly reversible should be viewed with caution. It may not even be acceptable to promote intra-uterine devices, because of the difficulty of insertion through a nulliparous cervix, and the increased risk of pelvic inflammatory disease leading to subsequent infertility as a result of tubal occlusion.

The combined low-dose oral contraceptive pill would seem to be ideally suited for couples living in an established, stable sexual relationship; the health risks of the pill for this age group are miniscule. This is particularly true in a developing country, where the risks are far outweighed by the benefits (see Chapters 3 and 4). But for those women who experience only the occasional unpremeditated and unforeseen act of intercourse, using the Pill is like using a sledgehammer to crack a nut. Coitus interruptus is one possible alternative solution, but this is particularly unreliable for sexually inexperienced young men who may achieve orgasm quickly and may not know how to anticipate the point of ejaculatory inevitability. Women would be better advised to encourage their partner to use a condom, which also protects against sexually transmitted disease, or else they themselves could use a diaphragm and a spermicidal cream. And if they are caught off-guard, they need access to a 'morning after' pill.

In 1968, Pope Paul VI published his famous encyclical on the regulation of birth, *Humanae Vitae*. In it, he said:

The direct interruption of the generative process already begun and,

Table 2.1. *Percentage of unmarried women in the United States aged between 15 and 19 who have already had sexual intercourse. Results broken down by age and colour for the years 1971 and 1976*

Woman's age (years)	1971			1976		
	Total	Black	White	Total	Black	White
15	14	31	11	18	38	14
16	21	46	17	25	53	23
17	27	59	22	41	68	36
18	37	63	32	45	74	44
19	47	76	39	55	84	49

(From M. Zelnik and J. F. Kantner. First pregnancies to women aged 15–19, 1976 and 1971. *Fam. Plann. Perspect.* **10**, 11–20 (1978).)

above all, direct abortion, even for therapeutic reasons, are to be absolutely excluded as lawful means of controlling the birth of children. Equally to be condemned, as the Magisterium of the Church has affirmed on various occasions, is direct sterilization, whether of the man or of the woman, whether permanent or temporary. Similarly excluded is any action, which either before, at the moment of, or after sexual intercourse, is specifically intended to prevent procreation – whether as an end or as a means.

The only glimmer of hope is a let-out clause which also states that:

The Church teaches that married people may take advantage of natural cycles immanent in the reproductive system and use their marriage at precisely those times that are infertile, and in this way control birth.

Pope John Paul II has subsequently given his unswerving support to the principles laid down in *Humanae Vitae*, and so this poses a real problem for the world's 800 million Catholics. Many couples in developed countries are prepared to ignore his prohibition of contraception, but there is no doubt that in developing countries like the Philippines, or throughout large areas of South America, many still regard all artificial forms of contraception as sinful.

When Pope Paul spoke about 'natural cycles', he probably did not realize that the most important natural cycle of all is the extended birth spacing that can be achieved by breast feeding. The Catholic Church could do so much to improve maternal and child health in developing countries if it would encourage prolonged lactation. Instead, it has concentrated on periodic sexual abstinence throughout the fertile phase of the menstrual cycle as the only permissible form of natural contraception. Unfortunately, humans are the most covert ovulators of all mammals, and so the drawback to all 'rhythm' methods of contraception is that they depend on prolonged periods of sexual abstinence of 10 days or more per cycle. This may be particularly difficult for young couples to observe, and they would no doubt prefer to call it 'Unnatural Family Planning'.

There are many variants of periodic sexual abstinence, from simple calendar methods that do not permit intercourse around mid-cycle, to more sophisticated variants where basal body temperature recording is used to determine when ovulation *has* taken place, to the sympto-thermal methods where ovulation is also *anticipated* by recording the viscosity of the vaginal mucus (see Book 3, Chapter 6, Second Edition). Vaginal mucus does indeed provide a good predictive index of impending ovulation, but since viable spermatozoa can survive in the female reproductive tract for at least 5 days, it is necessary to predict ovulation well in advance of the event if one is to be on the safe side. The use–effectiveness of all Natural Family Planning methods is low, and ovulation can be particularly difficult to predict in teenagers, because in the years immediately following the menarche, cycle lengths and hence the timing of ovulation can be extremely variable.

Although there will be many disappointments among those who attempt to use Natural Family Planning to postpone their first pregnancy, there can be no doubt that at the population level this method, if widely used and practised conscientiously, could have considerable demographic impact.

Sadly, there will always be those who take risks, whether out of ignorance, or moral or medical fear of contraception, or because contraceptives are not available when needed, or through sheer bravado. An unplanned pregnancy too early in life may ruin a future by interrupting an education, terminating an employment, or forcing a marriage (see Table 2.2). For such individuals, therapeutic abortion is an essential option. The earlier the abortion is performed, the less the morbidity and mortality, so it is callous to introduce unnecessary legislative and legalistic delays to the system. To deny abortion to those who so desperately seek it is an

Table 2.2. *Legal minimum ages of marriage for men and women in developing and developed countries*
Predominantly Catholic countries, where abortion is illegal, tend to have the youngest ages for women, favouring 'shotgun marriages' for pregnant teenagers. More permissive developed countries, where incidentally the age at puberty is lowest, have the latest ages at marriage for women, thereby favouring teenage abortions.

Country	Legal minimum age for marriage (years)	
	Men	Women
Algeria	18	16
Chile	14	12
Columbia	16	14
Denmark	18	18
Egypt	18	16
Finland	18	17
France	18	15
West Germany	21	16
Italy	16	16
Japan	18	16
Mexico	16	14
Norway	18	18
Spain	14	12
Sweden	18	18
Switzerland	20	18
Turkey	17	15
United Kingdom	16	16
Venezuela	14	12
Yugoslavia	18	18

(From L. Engstrom. Teenage pregnancy in developing countries. *J. Biosoc. Sci.* Suppl. **5**, 117–26 (1978).)

unwarranted restriction of the woman's freedom; to force her to attend a lay abortionist using antiquated and hazardous procedures (see Chapter 3) is to put her life in danger. And for a developed country like the United States, that endorses legal abortion for its own people, to refuse aid to a developing country for abortion on moralistic and political grounds is not only hypocritical, but smacks of intellectual imperialism.

Contraception to achieve adequate birth spacing

From a demographic standpoint, there is no doubt that the spacing between births is a much more important determinant of population growth rates than the age at first birth (see Book 4, Chapter 2, Second Edition). The only possible exception is China, whose one-child-family policy would make birth spacing an irrelevance!

A recent report by the Director of UNICEF (the United Nations Children's Fund) has clearly documented the enormous impact of increased birth spacing for improving maternal and child health in developing countries (see Fig. 2.6). Their slogan of 'Too close, too many, too young' says it all. Some women in developing countries may intuitively realize this, but if the means for increasing the spacing between births is simply not available, how can they be expected to respond? This is an excellent example of how the provision of appropriate contraceptives could dramatically improve infant health and survival; furthermore, if women could be sure that the children that they do have are not going to die, they would be much more prepared to restrict the size of their families. When we in the Western world look at the question of contraceptive safety, we think only in terms of the morbidity and mortality rates for individuals using the contraceptive, with never a thought for the enormous indirect benefits that might accrue in a developing country as a result of increased birth spacing. UNICEF studies in India, Turkey, the Philippines and Lebanon, for example, have shown that infant mortality rates for babies born within 1 year of a previous birth are between *two and four times* as high as for babies born after an interval of 2 years or more.

The current debate in the United States about the licensing of the injectable contraceptive Depo-Provera illustrates this very point. This steroid, a gestagen, is given by injection once every 3 months, and is an extremely effective contraceptive which acts by inhibiting ovarian follicular development and ovulation. It is licensed for contraceptive use in many developed (e.g. United Kingdom, Sweden, Switzerland, West Germany, New Zealand) and developing countries, and has already been used by an estimated 11 million women, at least half of whom have come from developed countries. There is not a *single* recorded case of maternal or infant mortality ascribable to the use of this drug. World Health Organization studies published in 1984 have provided no evidence to suggest that its use might increase the risk of breast cancer, and they have shown that it could be ideal for use in breastfeeding mothers to increase birth spacing, since unlike the combined oral contraceptive pill, it has no adverse effect on milk volume or composition. It also induces a state of amenorrhoea, and hence would be ideal for extending the normal period of lactational amenorrhoea.

Unfortunately, Depo-Provera has not received United States Food and Drug Administration (FDA) approval for contraceptive use within the United States, and according to Federal law this means that government funds cannot be used to promote its use as a contraceptive in developing countries. In response to an appeal by the manufacturers, the Upjohn

Fig. 2.6. Too close, too many, too young. The adverse effects of short birth intervals, large family size and young maternal age on infant survival rates in developing countries. (Source: *People*, **11** (1), 27 (1984).)

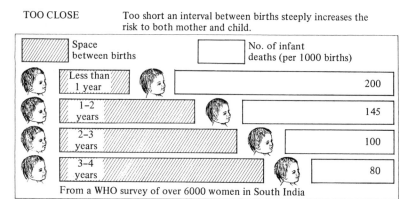

TOO CLOSE — Too short an interval between births steeply increases the risk to both mother and child.

From a WHO survey of over 6000 women in South India

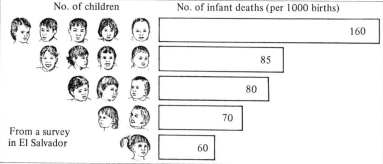

TOO MANY — The risks to the health of both mother and infant increase steeply after the fourth child.

From a survey in El Salvador

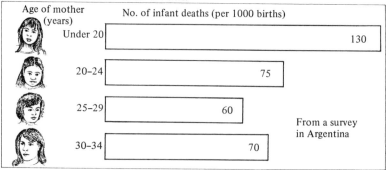

TOO YOUNG — Children born to women under the age of 20 are approximately twice as likely to die in infancy as children born to women in their mid-twenties.

From a survey in Argentina

Company, the FDA agreed to set up a public Board of Enquiry to review the evidence. Their first term of reference was to determine:

> Whether, in comparison with other drugs approved for contraception, the benefits of Depo-Provera in the United States outweigh its risks under conditions of general marketing.

The inclusion of those words 'in the United States' at once excludes from consideration all the evidence from those areas of the world where the drug would have its most beneficial effects in reducing infant and child mortality through increased birth spacing. In the United States, there is probably little need for Depo-Provera as an additional contraceptive. The birth rate has already fallen almost to replacement level thanks to the availability of excellent contraceptive and abortion services, so there is no call for increased birth spacing. Even if there was, perinatal and infant mortality rates are already so low that there would be little additional benefit to be gained. And in a country where few women choose to breast feed their babies for any length of time, there is no demand for a contraceptive that facilitates prolonged lactation.

The Board of Enquiry met on numerous occasions, received volumes of written and oral testimony – in September 1983, when they began, they commented that the documents in the administrative record occupied 45 linear feet of shelf space; by the time they submitted their report on 17 October 1984, this had increased to 54 feet! – and concluded that the evidence submitted (11 million women, hundreds of dogs, dozens of monkeys) 'does not provide sufficient basis from which FDA can determine that DMPA [Depo-Provera] is safe for general marketing in the United States'.

So there the matter stands. Guilty until proven innocent. Another case of beams and motes?

Depo-Provera is the forerunner of a whole new generation of improved gestagen-only injections and implants based on the steroids norethisterone enanthate or laevonorgestrel (see Fig. 2.7). These are already under extensive clinical trial in developing and developed countries around the world, and look extremely promising (see below).

Since adequate birth spacing is such a crucial determinant not only of lifetime fertility, but also of maternal and infant health, what strategies can be developed to maximize this spacing? Immediately, we run into an organizational obstacle. In the Health Departments of most developed and developing countries, and even within the World Health Organization itself, separate departments deal with family planning, and with maternal and child health. But if we are to maximize birth spacing, this demands inputs from both these areas, and a completely integrated approach.

Let me illustrate what can go wrong if there is no co-ordination. It is generally believed by demographers that in developing countries today, with the exception of China, breast feeding has a greater contraceptive impact than all modern forms of contraception put together. In 1984, for example, the World Bank estimated that about 60 per cent of all couples in the developing world were not using any modern form of contraception, a figure that ranged from 30 per cent in Singapore to over 90 per cent in most of Africa. In the absence of modern contraceptives, mankind is therefore thrown back on Nature's traditional birth spacer, breast feeding (see Book 4, Chapter 2, Second Edition). Anything that undermines breast feeding can therefore have a *conceptive* effect; breast milk substitutes, artificial feeding bottles, 'dummies' or 'comforters', even maybe thumb sucking, can all be expected to stimulate fertility in a non-contracepting population (see Fig. 2.8), because they will reduce the frequency with which the baby sucks from the breast, thereby interfering with the neural reflex responsible for the inhibition of ovulation (see Book 3, Chapters 6 and 8, Second Edition). Thus, when maternal and child health authorities advocate the early introduction of supplements of any sort into the baby's diet, they may not realize that they are inadvertently stimulating the mother's fertility (see Fig. 2.9). Unfortunately, the recent World Fertility Survey has shown that in developing countries today it is the young, urban,

Fig. 2.7. Structural formulae of progesterone and three synthetic gestagens.

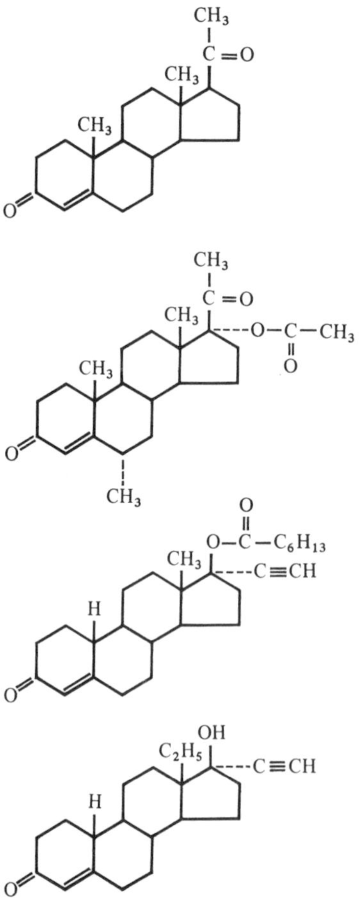

Progesterone

Depo-Provera
alias
Depo-medroxyprogesterone
acetate (DMPA)
alias
6α-methyl-17-acetoxyprogesterone

Norethisterone
enanthate (NET-EN)

Laevonorgestrel

educated women, the trend-setters in the community, who are the ones who are abandoning breast feeding in favour of the bottle. In Kenya, for example, which now has the second highest rate of population growth in the world and only 7 per cent contraceptive usage, much of the recent sharp rise in fertility to the present average level of about eight births per woman

Fig. 2.8. Conceptives which increase fertility by eroding the natural contraceptive effect of repeated nipple stimulation. These include feeding bottles, powdered milk, 'dummies', thumb sucking, and the combined, low-oestrogen oral contraceptive pill which decreases milk yield, leading to premature weaning.

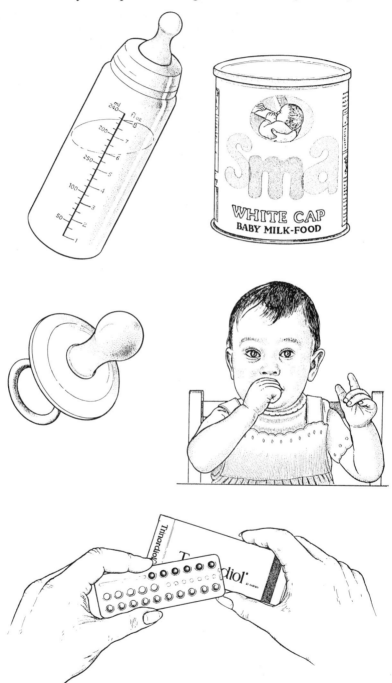

Fig. 2.9. An Australian
aboriginal woman, 20 years
old, with her six children.
Her high fertility is a direct
consequence of a
government-sponsored
scheme to provide
powdered milk to the
mission station and
encourage early weaning,
while failing to provide
alternative contraceptive
support. All these children,
although of normal weight
at birth, received a severe
growth check at weaning,
often not gaining weight
for a year or more, and
never showing complete
catch-up growth thereafter.
(From J. W. Cox. Effect of
supplementary feeding on
infant growth in an
Aboriginal family. *J.
Biosoc. Sci.* **10**, 429–36,
Plate 1, Fig. 1 (1978).)

Fig. 2.10. The changing
pattern of breast feeding in
Kenya today. World
Fertility Survey data to
show how maternal age
(15–24; 25–34; 35–49),
level of education
(NS = no schooling;
P = primary education;
S = secondary education)
and an urban or rural
environment influence the
duration of breast feeding.
The elderly, uneducated,
rural population, leading a
traditional lifestyle, breast
feed for longest, while the
young, educated, urban
population are leading the
trend towards earlier and
earlier weaning.

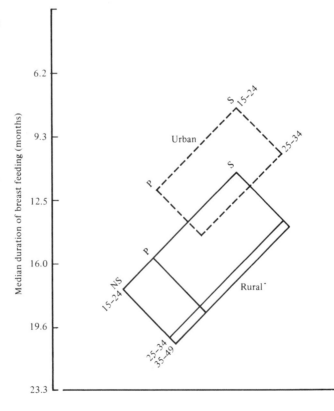

can be attributed directly to a decline in the duration of breast feeding (see Fig. 2.10). In Bangladesh, where the mean duration of lactational amenorrhoea is 18.5 months and only 9 per cent of women use contraceptives, contraceptive usage would have to rise to 43 per cent just to hold fertility at its present level, if lactational amenorrhoea were to decline to 6 months.

Not only can injudicious maternal and child health programmes set back family planning, but the converse may occur. It came as something of an unpleasant surprise to discover that the provision of the low-dose combined oral contraceptive pill to Bangladeshi women early in lactation paradoxically *stimulated* their fertility; they became pregnant sooner than those who were not given the Pill and allowed to continue breast feeding. We now know that the combination pill inhibits lactation, which is certainly not desirable from the baby's point of view. The most likely explanation for the increased fertility is therefore that those on the Pill weaned their babies early as a result of a diminished milk supply, thus prematurely losing the contraceptive effect of breast feeding. But the Pill is not popular in Bangladesh, and continuation rates are notoriously low – about 6 months of use on average. So the Pill-takers, having lost the natural contraceptive effect of lactation, promptly abandoned the artificial contraceptive and found themselves pregnant again all too soon.

It is a matter of historical record that the incidence of breast feeding in most developed countries has been on the decline for a long time. Royalty and the aristocracy set the fashion as far back as the sixteenth and seventeenth centuries by sending their children out to wet nurses, which had a staggering impact on the number of children born. Consider the case of Mary, wife of the fourth Earl of Traquair, who lived in Scotland, just south of Edinburgh, from 1670–1759 and produced 16 children (Fig. 2.11). Royalty were also among the first to adopt formulae based on flour and cow's milk for the artificial feeding of their offspring; Queen Victoria's large family of nine children was probably a consequence of this trend. As it permeated down through society to the working classes, who were now streaming into the towns and cities in the wake of the industrial revolution (Fig. 2.12), childhood mortality reached alarming proportions (see Table 2.3). The main cause of this mortality was diarrhoea, produced by a combination of insanitary surroundings, an inappropriate diet, and

Fig. 2.11. Dates of birth of the children of Mary, wife of the fourth Earl of Traquair. She was married at the age of 24, and gave birth to her first child a year later. All her children were sent out to wet nurses, and as a result she succeeded in producing 16 children in the space of 17 years, including one set of twins. (Source: plaque in Traquair House, Peebles, Scotland.)

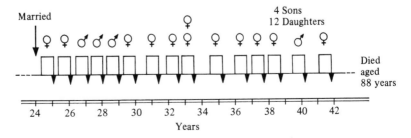

Fig. 2.12. An early flash photograph of a mother and child in a Glasgow slum, c. 1910. Note the feeding bottle warming on the stove, an ideal way of encouraging bacterial growth which will result in infant diarrhoea and growth retardation. (Source: *The Oxford Illustrated History of Britain*. Ed. K. O. Morgan, p. 516. Oxford University Press (1984).)

Fig. 2.13. Recent trends in the percentage of mothers attending Infant Welfare Clinics in the State of Victoria, Australia, who were still breast feeding their babies at 3 and 6 months of age (1950–80). This is exactly the reverse of the trend in most developing countries. (Source: Health Commission of Victoria, Annual Reports of Maternal and Infant Health (1950–80).)

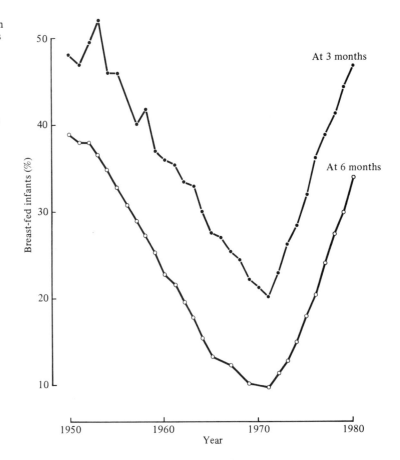

lack of the immunoglobulins secreted in breast milk which normally protect the infant against any enteric pathogen to which the mother has already acquired an immunity (see Book 4, Chapter 6, Second Edition). It is only within the last 10 years or so that we have seen a resurgence of interest in breast feeding by the young, urban, educated women of developed countries (Fig. 2.13) – ironically the very group who are leading the trend *away* from breast feeding in the developing countries!

It is perhaps not surprising that in such a climate of opinion, family planning organizations in developed countries have been particularly slow to appreciate the enormous importance of breast feeding for birth spacing in a developing country; the World Health Organization's Expanded Programme of Research, Development and Research Training in Human Reproduction, for example, has only started to fund research in this field within the last year. Another reflection of this bias comes out in a WHO survey of patterns and perceptions of menstruation – the largest international cross-cultural survey of its kind ever undertaken. Information was obtained from 5322 women in 14 different socio-cultural groups in 10 countries, namely Egypt, India, Indonesia, Jamaica, South Korea, Mexico, Pakistan, the Philippines, England and Yugoslavia. Among other things, the women were asked whether or not they would accept a method of contraception which would result in cessation of their menstrual periods. The general conclusion was that the majority of women in all cultures

Table 2.3. *Records of births, and deaths of children under the ages of 2 and 5, in the City of London between 1762 and 1771*
These figures are taken from Hugh Smith's treatise on *Nursing and the Management of Children*, published in 1792; he ascribed this staggering mortality principally to 'The thrush and watery gripes... both of them totally occasioned by improper food; such as all kinds of pap, whether made from flour, bread or biscuit: they all cause too much fermentation in an infant's stomach, and irritate their tender bowels beyond what Nature can support.' He concluded, 'Let me then intreat those who are desirous of rearing their children, not to rob them of their natural breast.' He was so right.

Year	Total births	Burials under 2 years of age	Burials under 5 years of age
1762	15351	8372	10659
1763	15133	8200	11163
1764	16801	7673	9699
1765	16374	8073	9948
1766	16257	8035	10197
1767	15980	7668	9449
1768	16042	8229	10670
1769	16714	8016	10061
1770	17109	7994	10121
1771	17072	7617	9447

investigated were not prepared to accept induced amenorrhoea, one of the main reasons being a fear that their general health might be impaired. This conclusion has had an important influence on all the WHO's subsequent strategies for contraceptive development – they feel that any method that suppresses menstruation would automatically be culturally unacceptable to the majority of women in developing countries.

Such a conclusion may be wrong. If we look at the way in which the WHO survey was conducted, we find that in order to be included, a woman had to be parous, non-pregnant, non-menopausal and non-lactating at the time of the interview. In other words, she would almost certainly have been having a succession of menstrual cycles, presumably because she or her partner was using some form of contraception, or because she was trying to get pregnant again. Such women are hard to find in a developing country. Indeed, the investigators themselves noted that 'one difficulty in Mexico was that most women resident in the rural areas selected tended to breast feed for very long periods after a birth. This made a large proportion of them ineligible'; they encountered similar problems in India. Thus, the WHO sample was highly biased, to the point of excluding the majority of the fertile female population. How much more informative the survey would have been if they had questioned women in lactational amenorrhoea, and asked them if they would welcome the use of a contraceptive that did not adversely affect their milk yield or the health of their baby, but which would postpone the resumption of their menstrual periods, and the return of their fertility. From enquiries that I have made in Egypt, India and Australia, I am convinced that such a survey would produce a very different answer; most breast-feeding women would welcome an extension of their lactational amenorrhoea.

When considering the question of the most appropriate form of contraception for a breast-feeding woman, we need to avoid falling into the trap of inadvertently providing 'double cover'. There is no point in giving a contraceptive to a woman immediately post-partum, when she will already be fully protected by breast feeding. So when does additional contraceptive protection first become necessary? To answer that question, we must probe more deeply into the physiology of lactational amenorrhoea; unfortunately, the only adequate data available are from studies performed in developed countries. These may not be strictly applicable to developing countries, because of differences in the plane of nutrition which may have an important bearing on the problem, a point to which we will return later.

Studies carried out by Peter Howie and his colleagues in Edinburgh, and Jim Brown and his colleagues in Melbourne, have shown that 58 per cent of breast-feeding women resumed menstruation without a preceding ovulation, the menstruation in such cases being an oestrogen-withdrawal effect as one of the developing follicles in the ovary, having failed to achieve ovulation, became atretic and ceased oestrogen secretion. Even when ovulation preceded the first post-partum menstruation, it was often followed by a deficient luteal phase, when the corpus luteum secreted an abnormally low amount of progesterone for an abnormally short period of time; such cycles are thought to be incompatible with the establishment of a normal pregnancy. Jim Brown has shown that only 19 per cent of the women in his study had a normal ovulatory cycle preceding the first menstruation. Since the maximum probability of conception in a normal menstrual cycle is only about 25 per cent (see Book 4, Chapter 2, Second Edition), this suggests that only about 5 per cent of lactating women who resume unprotected intercourse before the first post-partum menstruation are likely to become pregnant. This is in good agreement with the observed figure of 2–8 per cent of women in developing countries who become pregnant while breast feeding before their menstrual cycles have resumed.

Although a 2–8 per cent failure rate of lactational amenorrhoea as a contraceptive might be totally unacceptable to a woman in a developed country, who expects to be able to plan her family with clockwork precision, in a developing country setting this is no higher than the failure rate with the best of modern contraceptives, which are often not used correctly, and which have high discontinuation rates. Thus, from a demographic viewpoint, it would be acceptable to defer the provision of alternative contraceptives until the time of the first post-partum menstruation, which serves as a clear warning to the woman that her fertility is about to return. Alternatively, if a greater degree of contraceptive protection is needed, it should be a relatively simple matter, knowing the breast-feeding practices of a particular community, to anticipate the probable time of resumption of menstruation, and to provide contraceptive protection prior to this event.

For the woman who wants to continue to breast feed for as long as possible, and to extend the duration of lactational amenorrhoea, gestagen-only contraceptives are the answer. The gestagen-only minipill, taken daily by mouth, is one solution, although unfortunately it is still not readily available in most developing countries. Three-monthly Depo-Provera injections are also ideally suited to the breast-feeding mother, who may prefer an infrequent injection to regular pill-taking. It is to be hoped that the new gestagen implants and injections will have a major contribution to make at this particular stage of a woman's reproductive life history (see below).

Before leaving the topic of breast feeding and birth spacing, there are some nutritional aspects to discuss that are all too often neglected in family planning programmes. One of the common diseases of childhood in developing countries is kwashiorkor, a West African term meaning literally 'the evil eye of the child in the womb upon the child already born', a prophetic description of the effects on the older child if the mother becomes pregnant again too soon, thereby prematurely cutting off her milk supply. Thus, kwashiorkor, a consequence of severe malnutrition, can apparently be caused by birth intervals that are too short. It occurs in those regions of the world where children are already on a marginal plane of nutrition, and hence may be critically dependent on the proteins and calories normally supplied by their mother's milk during the first 2 or 3 years of life. The signs of kwashiorkor are not a pleasant sight to behold (see Fig. 2.14). Sadly, they are all too familiar to those of us who have viewed the horrors of the recent Ethiopian famine on our television screens, sitting in the obscene luxury of our homes. The children all have characteristic pot bellies as a result of gross enlargement of the liver produced by fatty infiltration, often accompanied by cirrhosis. The hair on the scalp is scanty, straight, silky and depigmented, giving it a bleached appearance often with a reddish tinge. Apart from their pot bellies and oedema of the legs, the children appear grossly emaciated; depigmentation of the skin gives it a crazy-paving appearance, with numerous deep necrotic ulcers and sometimes gangrene of the limbs. There is constant diarrhoea with fatty, semi-fluid, evil-smelling stools. Such children are dull, apathetic, have lost all desire to eat, and sit huddled against the cold, even in hot weather, uttering pathetic, miserable, low whimperings. They are particularly susceptible to any disease that is around, and the mortality rate is appallingly high. Many of the pathological changes are irreversible, so that death is often inevitable even after aid has arrived.

Famine has stalked the land from time immemorial, and will continue to do so with increasing frequency in the future as more and more people are born in those regions of the world that are less able to support them. It seems incredible that at this late stage we still lack precise scientific information on something as simple as whether or not we can increase a malnourished woman's milk yield by giving her more food. Neither do we

Fig. 2.14. Indonesian children with kwashiorkor, all 3 years of age. Only the first child (on the left) was able to stand upright, and he is supporting the others. All show ascites, big livers, oedema and extensive muscle wasting. (Source: World Health Organization.)

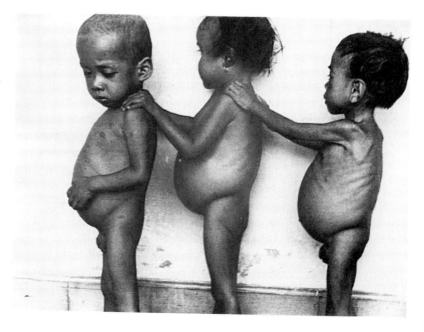

have any clear guidelines about when and how best to start introducing supplements into the diet of the breast-fed baby. Of one thing we can be quite certain: nutrition, lactation and reproduction are inextricably linked to one another, and of enormous importance in determining birth spacing and hence completed family size in developing countries. If kwashiorkor should indeed prove to be a disease of short birth intervals, appropriate contraceptives might prove far more effective in preventing it than countless shiploads of grain. To export powdered milk to developing countries is one of the worst things a developed country can do, because it will only stimulate fertility by undermining traditional breast-feeding practices. How unfortunate that in 1981 the United States, the home of some of the world's largest powdered milk manufacturers, was the only nation to refuse to endorse the World Health Organization Code of Practice that sought to limit the aggressive advertising and sales of breast milk substitutes in developing countries, on the grounds that the code was an unwarranted intrusion into a country's internal affairs.

Contraception after attainment of desired family size
Readers of Malcolm Potts's chapter (Chapter 3) will appreciate that many women in developing countries have more children than they actually want (see Fig. 3.7), so the question of how to curtail fertility when the desired family size has been achieved is an important one. The solution adopted will depend on the particular culture. In countries where the divorce rate is high, or where infant mortality is high, some couples may be reluctant to accept irreversible male or female sterilization, especially when they are still young. On the other hand, we know that the health risks from continued use of the oral contraceptive pill begin to increase appreciably after a woman reaches the age of 35, particularly if she smokes (see Chapter 4). As women approach the menopause, their menstrual cycles become increasingly irregular, making it more difficult to rely on 'rhythm' methods of periodic sexual abstinence. Menstrual flows also tend to become heavier, and this would only be exacerbated by the presence of an intra-uterine device. We know that up to 50 per cent of women in some developed countries undergo a hysterectomy at this stage, one of the principal indications being for disorders of menstruation; although hysterectomy might be said to represent the ultimate in female contraception, it is hardly a practical solution for a developing country, desperately short of skilled medical manpower and hospital facilities.

Another factor to be taken into account is that an unexpected pregnancy towards the end of a woman's reproductive life is not only socially disruptive, but it presents the greatest obstetric hazard for the mother, who also has an increased chance of producing a handicapped child (see Book 2, Chapter 5, Second Edition). Thus, whatever form of contraception is adopted needs to be highly reliable. Let us consider some of the options in turn, with their advantages and disadvantages.

Male sterilization

Although vasectomy has a clean bill of health, despite some recent scares (see Chapter 4, and also Book 4, Chapter 6, Second Edition), it still has a number of drawbacks. The first is its relative irreversibility. While it may be possible to re-anastomose the cut ends of the vas, this requires a long operation and skilled microsurgery, and even then there can be no guarantee that fertility will be restored. Such facilities are usually not available in developing countries.

The technique for vasectomy is simple enough, and can be performed under a local anaesthetic as an outpatient procedure. It is therefore amazing to discover the number of doctors in developed countries who still insist on carrying it out under a general anaesthetic; the risks of the anaesthesia far outweigh those of the procedure itself. Perhaps this ultra-cautious approach is to avoid blow-by-blow accounts which appear in the press from time to time, like John Payne's memorable article in the *Guardian* in 1971:

> Belatedly the surgeon came over and draped me with a green tablecloth sort of thing. As he shook it out I saw there was a slit in the middle, and it didn't take much working out what this was for. Lying flat out as I was I couldn't see what he was up to down there, but I felt him draw me through the slit and arrange me on top of the cloth.
>
> God, I thought, now I must look like sweetbreads on green salad.
>
> The needle stuck me, and in spite of my toes being curled up tight inside my shoes my right leg jerked as if I'd been plugged into the light socket. But the anaesthetic took effect immediately and after that he could have been using a blunt breadsaw on me for all I knew. All I felt was ridiculous.
>
> The surgeon gave me one of his relax-and-enjoy-it smiles. 'Not hurting you at all, am I?' he asked.
>
> 'No, not a bit,' I had to admit. (Then why was I sweating so much?). 'The only thing bothering me is what it's going to be like once the anaesthetic's worn off.'
>
> 'Oh', he said airily, 'most people tell me it's no worse than being hit there by a cricket ball.'
>
> I didn't like to say that that had always been the worst possible thing I could imagine happening to me. ⌄

It is perhaps no wonder that vasectomy has not proved particularly popular in chauvinistic, male-dominated societies where a man and his machismo are not easily parted. For example, the operation is rarely performed in South America.

Drs Li Shun-quiang in Chungking and Wu Chieh-ping in Beijing, China, have recently pioneered an exciting new development in male sterilization by developing a simple non-surgical injection–occlusion procedure that does not even require a skin incision. Their technique is as follows: a bleb of local anaesthetic is injected into the scrotal skin over the top of the vas on either side and, following puncture of the vas, a fine cannula is then inserted into its lumen and the tract is flushed centrally with a dye, using methylene blue on the right side and rivanol (yellow) on the left. The patient is then asked to pass some urine; judging from its colour (blue, yellow or green), it is easy to tell when both cannulae are correctly in place, and a small volume of butyl cyanoacrylate containing phenol is then injected into the lumen of the vas on either side to produce a permanent occlusion. Alternatively, if it is desired to make the technique reversible, Dr Zhao Sheng-cai from Shanxi has shown that a plug of polyurethane can be used. This can simply be 'shelled out' of the vas at some later date by means of a small longitudinal incision which can easily be closed with a suture.

These techniques are brilliant in their simplicity; with characteristic humility, the Chinese say that as yet they have only performed these procedures on a mere 500000 men!

Another potentially exciting discovery from China concerns the possible use of gossypol as a male contraceptive (see Fig. 2.15). The chemical structure of gossypol, first identified in cottonseed in 1899, was finally established in 1938. It is well known that when given in large doses to domestic and laboratory animals, gossypol is highly toxic, causing pulmonary and cardiac oedema, diarrhoea and paralysis. Cottonseed cake is widely used throughout the world for livestock feed, and cottonseed oil is used as a cooking and salad oil, and in margarine. Fortunately, cottonseed is readily detoxified by heating, which conjugates the gossypol to proteins, thereby inactivating it.

In the 1950s, outbreaks of infertility were reported in a number of rural communes in China, and it seemed probable that the use of crude cottonseed oil for cooking might be responsible. For years, Chinese farmers had processed cottonseed oil for human consumption by first heating the seed, and then pressing the oil out. Then, during the 1950s, the farmers started to take their cottonseed to a central location where it was pressed without heating. After using this oil for cooking for a year or more, the men became infertile and the women amenorrhoeic. When use of the oil was discontinued, the women started to menstruate again, but some of the men did not regain their fertility. In 1972, the Chinese

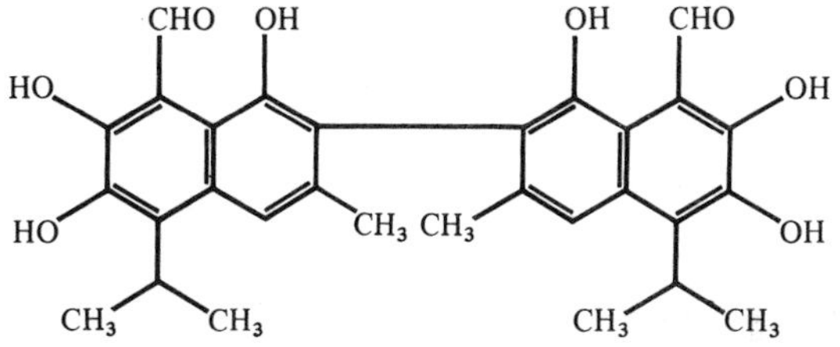

Fig. 2.15. Gossypol, a male anti-fertility agent present in unprocessed cottonseed oil.

established a National Co-ordinating Group on the Male Anti-fertility Agent Gossypol, and clinical trials of gossypol acetic acid were initiated. Men were given a daily dose of 20 mg for about 2 months, followed by a maintenance dose of 75–100 mg twice a month once necrospermia (all spermatozoa dead) or oligospermia (less than 4 million spermatozoa/ml ejaculate) had been achieved. The results of the first 8806 men who had taken gossypol acetic acid for between 6 months and 4 years were published in the *Chinese Medical Journal* in 1978. Gossypol appeared to be 99.9 per cent effective as a male contraceptive, producing complete azoospermia with prolonged treatment, but without any adverse effects on the serum LH or testosterone levels. After cessation of medication, some men become fertile again after 3 months and subsequently produced normal, healthy children, while others continued to remain azoospermic. This, gossypol treatment could be likened to a partially reversible 'chemical vasectomy'.

However, gossypol treatment produced some adverse side-effects, which included changes in the electrocardiograph and a feeling of muscular fatigue. In Nanjing, there was a 4.7 per cent incidence of hypokalaemic paralysis, but this high incidence was not initially noticed in other areas of China. It is now estimated that about 10000 Chinese men have taken gossypol, and clinical follow-up of these individuals is continuing, especially those who have shown signs of paralysis. Unfortunately, it seems that the hypokalaemia is due to permanent damage to the proximal and distal convoluted tubules of the kidney, and that hypokalaemia cannot be over-come by increasing the dietary K intake.

Gossypol has now become the subject of much research in the West, and the World Health Organization has initiated an active research programme on the compound. Gossypol acetic acid is present as a racemic mixture of the (+) and (−) isomers, and so the obvious hope was that it might be possible to separate the desired contraceptive effect from the undesirable hypokalaemic effect. Chemists in Beijing and London have now succeeded in preparing pure (−) gossypol, which has twice the anti-fertility activity of the racemic mixture when tested in laboratory animals, and the isomer is now undergoing toxicity testing in animals. Alas, preliminary toxicology results are disappointing, and at the end of the day gossypol will probably prove just too toxic for clinical use. So near, and yet so far; there is no doubt that 'chemical vasectomy' would be an enormously important addition to the list of contraceptive agents, particularly since hitherto we have had no male methods other than condoms and surgical vasectomy.

Female sterilization

The classical Western techniques for tubal ligation are well illustrated in Martin Vessey's chapter (Chapter 4). But are they too sophisticated for use in a developing country? They certainly require hospitalization, a general anaesthetic, and a trained surgeon, which at once puts them out of reach of many of the people most in need. Attempts to occlude the Fallopian tubes by non-surgical means, such as the instillation of quinacrine or even superglue, have not been particularly successful, and require more than one administration for 100 per cent effectiveness. However, as Malcolm Potts points out (Chapter 3), there is no doubt that there is an enormous unmet demand for a simple, non-surgical female sterilization procedure in developing countries. It is interesting to see that, in China, tubal ligation performed by a mini-laparotomy is becoming increasingly popular, and leading to a marked decline in the use of intra-uterine devices.

The key question we should be asking ourselves when trying to formulate the optimal contraceptive strategy for a woman who does not want to have any more children is as follows: what is the optimal state of health for a breast that no longer needs to lactate, a uterus that no longer needs to menstruate, and an ovary that no longer needs to ovulate? As yet, we can only guess at the answers, but there is much to be said for trying to put the whole of the reproductive tract, including the breasts, to sleep once childbearing is completed. But even here, there are compromises to be made. Surgical removal of the ovaries, for example, will decrease the likelihood of developing cancer of the breast, endometrium and ovaries. But it will also induce a surgical menopause with all the attendant physical problems resulting from oestrogen withdrawal; these include vasomotor disturbances leading to 'hot flushes' and night sweats, a dry vagina resulting in painful intercourse, and, perhaps most important of all, calcium resorption from the long bones rendering them particularly prone to debilitating fractures, such as those of the neck of the femur. It seems that every intervention we make in the reproductive process has some associated disadvantages; all we can do is attempt to maximize the benefits while minimizing the risks.

Contraceptives today: the People's Republic of China
If we are to improve the family planning services of the developing countries, we need to look to China, rather than to any developed country, to see how it can best be done. China carried out her third national population census on 1 July 1982, and it showed a total population of 1 008 175 288, an 84 per cent increase over the numbers recorded at the end of 1949. Thus China is the most populous nation on earth, and fortunately she also has one of the most effective family planning programmes in existence.

In his report to the National People's Congress on 30 November 1982, Premier Zhao Ziyang had this to say about China's population:

> To attain our goal in population control [keep population within 1 200 million] is going to be an extremely important and strenuous task. The whole society must pay full attention to this problem. We must take effective measures and encourage late marriage, advocate one child for each couple, strictly control second births and resolutely

prevent additional births so as to control population growth. Otherwise, the execution of our national economic plan and the improvement of the people's living standards will be adversely affected. Persuasive education must be conducted among the people of the whole country, especially among the peasants, to change radically the feudal attitude of viewing sons as better than daughters and regarding more sons as a sign of good fortune. We must, in particular, protect infant girls and their mothers. A couple that has only one daughter and brings her up well deserves greater commendation, support and reward than a couple that has only one son. The whole society should resolutely condemn the criminal activities of female infanticide and maltreatment of the mothers, and the judicial departments should resolutely punish the offenders according to law.

This is not merely empty rhetoric; the policies are in fact working. In 1982, the State Family Planning Commission carried out a nationwide fertility survey of one person per thousand head of population in all regions except Taiwan and Tibet. Over a million people were questioned, including over 310000 women aged between 15 and 67; we are told that 99.99 per cent answered the questionnaire!

The results of this massive undertaking are of particular interest. Women of childbearing age (15–49) made up 25 per cent of the total population, and married women, 17 per cent. The average age of marriage in 1981 was 22.8. The average number of children born to a woman reached a peak in 1963 of 7.5, and has fallen steadily since then to reach 2.63 in 1981.

Of particular interest was the usage of contraceptives; 69 per cent of women were using some form of birth control. Of the users, 50.2 per cent had an intra-uterine device, 25.4 per cent had had a tubal ligation, in 10.0 per cent of cases the husband had had a vasectomy, 8.2 per cent used oral contraceptive pills, and in 2.0 per cent of cases the husbands used condoms. Of the 30 million pregnancies in a year, 8 million were terminated by abortion, and there were 11.2 million first births, 6 million second births, and 4.8 million third or subsequent births. The result of the one-child-family policy is that China now has 33 million one-child couples, of whom 14 million have pledged not to have a second child. Although female infanticide has attracted much adverse comment in the West, the sex ratio of the total population is 106.3 males per 100 females. This should be compared with the sex ratio at birth in Western countries of 105.5 males per 100 females; if female infanticide does occur, it cannot be very prevalent.

It is interesting to look in more detail at some of the contraceptives that are currently in use in China.

Intra-uterine devices

There is no other country in the world that has made such extensive use of the intra-uterine device for contraception. Each region has developed its own devices, often based on patterns already in use in developed countries (see Fig. 2.16). Thus, the stainless steel rings used in Shanghai, Tienjing and Beijing hark back to the early days when Grafenberg in Germany was using a ring of coiled silver wire – until arrested by the Nazis in 1937 because of this work. The Hangchou copper T is very similar to its Western counterpart, which is based on the discovery by Jaime Zipper of Chile in 1969 that copper wire, wound around the stem of a T-shaped plastic device, would increase its contraceptive effectiveness. Sichuan has developed its own version of the Dalkon Shield, which has now been withdrawn in the West because of the increased incidence of uterine

Fig. 2.16. A range of intra-uterine devices currently in use in the People's Republic of China. The stainless steel devices are the most popular.

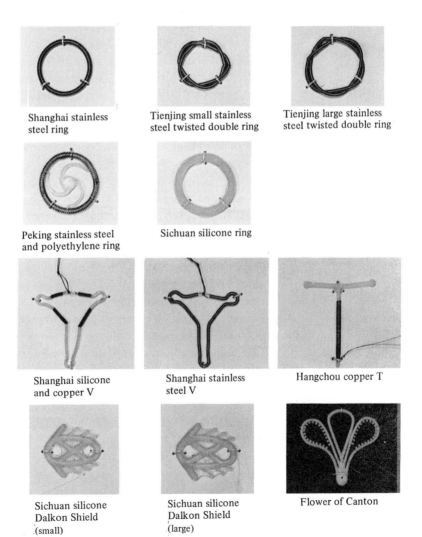

Shanghai stainless steel ring

Tienjing small stainless steel twisted double ring

Tienjing large stainless steel twisted double ring

Peking stainless steel and polyethylene ring

Sichuan silicone ring

Shanghai silicone and copper V

Shanghai stainless steel V

Hangchou copper T

Sichuan silicone Dalkon Shield (small)

Sichuan silicone Dalkon Shield (large)

Flower of Canton

infection and septic abortion associated with its use. But the Flower of Canton and the Shanghai V are uniquely Chinese devices.

There are several reasons for the popularity of the intra-uterine device in China: sexually transmitted diseases are rare, so there has been little problem with pelvic inflammatory disease; the device is cheap, and highly cost-effective; it is easy to train 'primary health care workers' in the rural areas to insert it, and once inserted the device gives good long-term contraceptive protection. However, recent epidemiological studies have shown that the stainless steel rings result in 10.6 pregnancies per 100 woman-years of use. This high failure rate accounts for a large number of the abortions that are performed, and, as newer and better IUDs are introduced, we can therefore expect to see a sharp fall in China's abortion rate.

Surgical sterilization

The fact that 25.4 per cent of contraceptive users in China opted for tubal ligation is striking testimony of the fact that it is possible to mount an effective female sterilization campaign, even in a developing country with a large rural population. Perhaps other countries should profit from the Chinese experience in this area. Vasectomy is much less popular, particularly in some regions, but even so an overall prevalence of 10 per cent is no mean achievement; perhaps the new percutaneous vas occlusion techniques already referred to will make the procedure even more popular in the years to come.

Oral contraceptives

Although fourth in order of popularity, the Chinese have developed a spectacular range of oral contraceptives. The main standbys are the Number 1 and Number 2 Contraceptive Pill, containing respectively 625 μg norethisterone $+35$ μg ethinyl oestradiol, and 1 mg megestrol acetate $+35$ μg ethinyl oestradiol, both types of pill to be taken daily for 22 days, starting on the fifth day of menstruation. These two formulations are also available as the Number 1 and Number 2 Paper Pill, where the steroids are impregnated into a perforated sheet of water-soluble carboxymethylcellulose paper, looking rather like a sheet of small postage stamps, one to be torn off and swallowed daily for 22 days. This ingenious, cheap and compact form of packaging, which also cuts down on the problems of steroidal dust contamination of workers in the production plant, has not proved as popular in practice as the conventional pill formulation.

China was quick to adopt a low oestrogen dosage of 35 μg ethinyl oestradiol in its two combination pills, 5 years before such low-dose pills were available in Europe or North America. Of the gestagens in the Chinese pills, norethisterone is also widely used in many oral contraceptives in the West; it was the first orally active synthetic gestagen, and was synthesized by Carl Djerassi and his colleagues in the Syntex laboratories in Mexico in 1951. Megestrol acetate was also in use in the West, until it was shown to cause mammary tumours in dogs and so was withdrawn by the licensing authorities. Since we now know that this is a feature of all gestagens, including progesterone itself, perhaps the Chinese were wise to continue using it.

The Chinese also have a Post-coital Pill, containing 550 μg megestrol acetate and 880 μg quingestanol (both gestagens), and a range of Visiting Pills for couples whose work forces them to live apart except for annual holidays of 2–4 weeks. Three of these visiting pills consist of a gestagen alone, to be taken daily for 14 days (2 mg megestrol acetate; 5 mg norethisterone; 3 mg norgestrel). A fourth consists of 80 mg of the gestagen quingestanol, to be taken once every 2 weeks, and a fifth, to be taken post-coitally, consists of 7.5 mg anordrin, a steroid that has oestrogenic activity and is also a weak anti-gestagen. Finally, they also have a Daily Pill, consisting of 300 μg norgestrel and 30 μg ethinyl oestradiol, and a Once-a-month Pill, containing 12 mg norgestrel and 3 mg quinestrol.

They also have a Monthly Injectable Contraceptive, in the form of 250 mg of the gestagen 17-hydroxyprogesterone caproate plus 5 mg of the oestrogen oestradiol valerate, but this has not proved very popular, presumably because of the disorders of menstrual bleeding that it would produce. About the only thing the Chinese seem to lack in the way of steroidal contraceptives is a gestagen-only minipill for daily use, or a gestagen-only implant or injection, or a gestagen-releasing intra-uterine device, all of which could be used for breast-feeding women.

Abortion

We cannot leave the subject of family planning in China today without making some mention of the very efficient abortion service. While developed countries were busy experimenting with laminaria tents to dilate the cervix, or intra-amniotic hypertonic saline to kill the fetus, or intra-amniotic, vaginal or systemic prostaglandins to induce uterine contractions, the Chinese went ahead and developed the simplest and most effective method of all, namely vacuum aspiration of the products of conception from the uterine cavity, a procedure that could be performed under traditional acupuncture anaesthesia. As Malcolm Potts points out (Chapter 3), this Chinese vacuum aspiration technique is now widely used around the world in developing and developed countries alike, and at the moment it still seems to be preferable to any of the medical abortifacients, since it has a higher success rate and a lower incidence of adverse side-effects such as nausea, vomiting and diarrhoea, which are a frequent complication of prostaglandin-induced abortion.

Tomorrow's contraceptives

Tomorrow's contraceptives are going to be very much the same as today's, with a few modifications principally with respect to dosages and delivery systems. Table 2.4 summarizes some of the medical advantages and disadvantages of various contraceptive agents. The challenge before us in designing tomorrow's contraceptives is to try to capitalize on these benefits while minimizing the risks.

Table 2.4. *The medical advantages and disadvantages of various contraceptive agents*

Advantages	Contraceptive	Disadvantages
Very effective central inhibitors of gonadotrophin secretion and hence ovulation. Can prevent heavy or prolonged menstrual bleeding by gestagen-primed endometrium. Prevent menopausal symptoms, e.g. hot flushes, dry vagina, osteoporosis	*Oestrogens*	Increased incidence of vascular disorders, e.g. thromboembolism, coronaries, stroke. Inhibit lactation. When used on their own, increase incidence of endometrial cancer and perhaps breast cancer
No known carcinogenic effects in humans. Do not inhibit lactation. Protect against carcinogenic effects of oestrogen on the endometrium and perhaps on the breast. Increase viscosity of cervical mucus, impeding sperm transport	*Gestagens*	Weak central inhibitors of gonadotrophin secretion, so waves of follicular development and atresia in ovaries result in irregular menstruation and breakthrough bleeding. Anabolic effects lead to weight gain
Increased surface area leads to increased contraceptive efficacy.	*Intra-uterine devices*	Increased surface area leads to increased blood loss at menstruation, and hence anaemia. Increased incidence of pelvic inflammatory disease, leading to subsequent infertility. Dysmenorrhoea

Steroidal contraceptives

There is no doubt that minimizing the dose of oestrogen in oestrogen–gestagen mixtures is highly desirable. But unfortunately, once oestrogen is removed altogether, there is a loss of cycle control, and irregular menstruation and breakthrough bleeding becomes common. This is apparently because of the relatively weak central inhibitory effects of gestagens on their own, thus allowing waves of follicular development and atresia to occur in the ovaries, with resultant irregular oestrogen-withdrawal bleeds from the gestagen-primed endometrium. Irregular or heavy bleeding can be controlled when it does occur by oestrogen administration, or by increasing the dosage of gestagen to achieve a greater degree of hypothalamic and pituitary inhibition.

It is interesting to compare the blood hormone levels achieved as a result of the infrequent injection of a long-acting gestagen (e.g. Depo-Provera, 150 mg every 3 months), or the daily oral administration of a gestagen-only minipill (e.g. norethisterone, 350 μg), or a slow-release gestagen implant (e.g. Norplant, see below) Fig. 2.17). There is no doubt that theoretically the implant is the best way of maintaining blood levels above the threshold necessary to inhibit ovulation, since it uses the lowest dose of steroid, and is effective over the longest period of time. However, we know that the large doses of Depo-Provera will produce amenorrhoea in 60 per cent of women after 1–2 years of administration, whereas Norplant implants will produce amenorrhoea in only 24 per cent of women after 9–12 months, and irregular bleeding in 12 per cent. If women prefer amenorrhoea to irregular and unpredictable bleeding, then a larger dose of gestagen may be the only solution.

There are now two new delivery systems for steroidal contraceptives that are already under extensive clinical trial, and we can be sure that these exciting new developments will soon become available for use in family planning programmes around the world.

Fig. 2.17. Diagram to illustrate the blood gestagen profiles following: A, injection of a long-acting gestagen; B, daily oral administration of a gestagen 'minipill'; C, subcutaneous implantation of a gestagen-releasing device. (Source: *Population Reports*, Series K, No. 2, Fig. 1 (May 1983).)

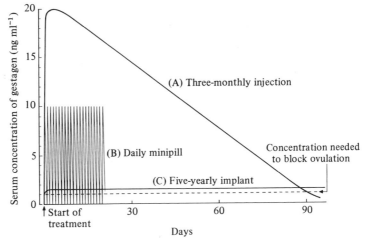

Norplant. This has been developed by the Population Council in New York, and exists in two forms. Norplant I consists of six hollow silastic tubes, each filled with 36 mg of crystalline laevonorgestrel and sealed at the ends. The capsules are inserted subcutaneously under local anaesthetic, usually beneath the skin of the forearm (see Fig. 2.18). The steroid slowly diffuses out through the wall of the tube at the rate of about 40 μg per day at the end of the first year, and 30 μg per day by the end of the fifth year. Norplant II consists of two slightly larger solid rods of silastic, each impregnated with 70 mg laevonorgestrel, to give a daily release rate of about 50 μg (see Fig. 2.19). Both types of device give excellent contraceptive

Fig. 2.18. Technique for insertion of Norplant under local anaesthesia. (*a*) Norplant I and Norplant II with trochar and cannula used for subcutaneous insertion. (*b*) Cannula in place beneath skin of forearm, and Norplant being inserted under local anaesthesia. (Photographs by courtesy of Dr V. Odlind, University of Uppsala, Sweden.)

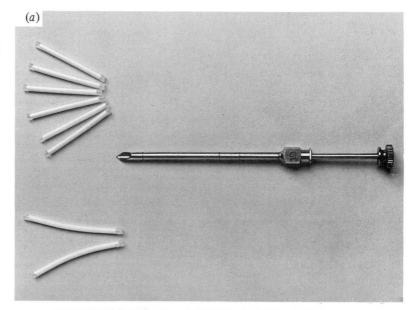

(*a*)

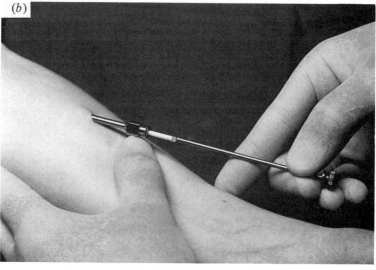

(*b*)

protection for 5 or more years. The effects seem to be readily reversible, since of 53 women who had the implant removed in order to conceive, 77 per cent had become pregnant within a year. The only disadvantage of Norplant is the problem of irregular menstruation in a proportion of users. But for reasons already stated, it would seem to be ideal for developing countries as an alternative to tubal ligation after the completion of childbearing, and also perhaps as a means of achieving adequate birth spacing without interfering with lactation. Norplant is already licensed for contraceptive use in Finland and Sweden, and it is hoped that other developed and developing countries will soon follow suit.

Vaginal rings. The Population Council have developed a steroid-impregnated silastic ring of 5.0 or 5.8 cm external diameter for inserting in the vagina around the cervix. Each ring contains about 100 mg of laevonorgestrel and 50 mg oestradiol-17β, and the average daily release rate from the rings is 250–290 μg of laevonorgestrel and 150–180 μg oestradiol; sufficient of the steroid is absorbed into the systemic circulation from the vaginal mucosa to inhibit ovulation for at least 6 months. The ring can be removed for a week to induce menstruation when so desired, and clinical trials have demonstrated that these rings are as effective in preventing pregnancy as low-dose oral contraceptives. However, the oestrogen dosage may prove to be unacceptably high.

The World Health Organization has opted for an alternative approach. Their vaginal ring (see Fig. 2.19), which is similar in size to the Population Council ring, contains only 6 mg laevonorgestrel. This is absorbed through the vaginal mucosa at the rate of about 20 μg per day for at least 90 days. The ring is left in place continuously, and does not interfere with the woman's normal ovulatory menstrual cycles; its contraceptive effect is

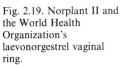

Fig. 2.19. Norplant II and the World Health Organization's laevonorgestrel vaginal ring.

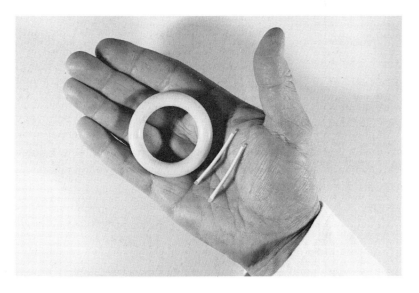

thought to be due, at least in part, to a local action on the viscosity of the cervical mucus, making it impenetrable to spermatozoa.

Although many women in developed countries already familiar with the use of cervical caps and diaphragms may welcome these new devices, it remains to be seen whether they will prove acceptable to women in developing countries. The relatively short life of the device is one obvious disadvantage, together with the fact that there is no visual indication of when the device has ceased to become effective.

Intra-uterine devices

While these have proved to be the mainstay of China's highly effective family planning programme, there is no doubt that the increased blood loss at menstruation and the increased risk of pelvic inflammatory disease are major drawbacks. There has therefore been much interest in trying to improve the design of the device. One ingenious attempt was to incorporate progesterone around the stem of a T-shaped intra-uterine device, in the hope that the hormone would have a local effect on the endometrium to cut down menstrual blood loss. This device is marketed as Progestasert. The stem of the device contains 38 mg of progestrone, which is released at a constant rate of 65 μg per day for a year. Ovulation is not affected, and blood loss and dysmenorrhoea (painful menstruation) is reduced. Unfortunately, the device has to be replaced each year, intermenstrual bleeding and spotting is a problem, and the ectopic pregnancy rate (implantation in the Fallopian tube) may be increased.

Dr Tapani Luukkainen from Helsinki, working with the Population Council in New York, has now developed one of the most exciting new steroid-releasing intra-uterine devices that seems to have overcome all the drawbacks of the Progestasert. Basically, it is a T-shaped plastic device with a silastic-coated stem that releases laevonorgestrel at a rate of 20 μg a day for up to 6 years (Fig. 2.20). The low systemic levels of gestagen are insufficient to interfere with normal cyclical ovarian activity, but within the uterine lumen the relatively high concentrations of laevonorgestrel saturate all the progesterone receptors, making the endometrium relatively insensitive to oestrogen. This results in a marked reduction in the number of days of menstrual bleeding per month to 1 or 2, with a mean monthly blood loss of only 20 ml; up to 30 per cent of women may experience amenorrhoea for a year or more. Since these gestagenic effects are purely local, oestrogenization of the vagina is unimpaired.

One of the most impressive things about this device is its high efficacy; the pregnancy rate was only 0.1 per 100 woman-years during the first 5 years of use, compared to 1.1–1.6 for a conventional copper intra-uterine device. Only one ectopic pregnancy has been encountered in 5000 woman-years of use. Normal menstrual cycles resume within about a month of removal of the device, and there is even evidence to suggest that the device might reduce the incidence of endometrial carcinoma, which is thought to be oestrogen-induced. It would certainly be an ideal form of therapy for excessive or prolonged menstrual periods, and thus might reduce the incidence of hysterectomy. It would also seem to be ideally suited to the needs of breast-feeding women, since it would probably prolong the duration of lactational amenorrhoea, and only miniscule quantities of the gestagen would enter the breast milk. Its ease of insertion, and ease of removal, are added advantages, and it may indeed prove to be more acceptable than Norplant. Leiras, the Finnish drug firm that manufactures it, is still carrying out clinical trials, but it really does look an exciting hope for the future that meets many of the criteria of a 'healthy contraceptive'.

Abortion

Although vacuum aspiration of the uterine cavity has proved to be a simple, highly effective way of terminating early pregnancies, it does require skilled staff and aseptic operating theatre conditions. Abortion is widely practised in developed and developing countries, and will always be necessary as a back-up for failures of contraception. In countries such as Japan, it still appears to be the principal method of fertility regulation, since in any given year there are more abortions than live births. This may be partly because the Japanese medical profession makes a large profit from abortions, and has a powerful political lobby to ensure that the oral contraceptive pill is banned on 'health' grounds. Be that as it may, it would be an enormous advance for developing and developed countries alike if abortion could become a simple outpatient or even perhaps a self-administration procedure.

That goal may now be within our grasp. A French pharmaceutical company, Roussel-Uclaf, has recently synthesized a novel orally active steroidal anti-gestagen, RU 38486 (see Fig. 2.21), which appears to act as

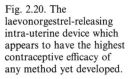

Fig. 2.20. The laevonorgestrel-releasing intra-uterine device which appears to have the highest contraceptive efficacy of any method yet developed.

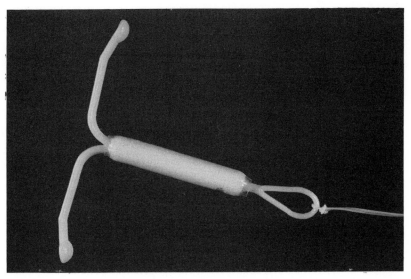

a competitive antagonist of progesterone by binding to its receptors in the target tissues. The West German drug company Schering has also synthesized similar anti-gestagens, and is actively involved in studying them. The World Health Organization carried out clinical trials of RU 38486 in Sweden and Hungary, giving it to a total of 36 women who were all less than 42 days pregnant and who wished to have their pregnancies terminated. The compound was given daily for 4 days in doses of 50–200 mg; all but three of the women started to bleed from the uterus during the course of the treatment, and 22 of the 36 subsequently had complete abortions. Encouraged by this partial success, the World Health Organization has now embarked on a clinical trial in which women seeking early termination of pregnancy are pre-treated with the anti-gestagen for 4 days, and are then given a single intramuscular injection of 250 μg of the synthetic prostaglandin analogue, sulprostone. Preliminary results are extremely encouraging, with complete abortions in all of 16 women with amenorrhoea of 49 days or less. The anti-gestagen pre-treatment increases the sensitivity of the uterine musculature to the prostaglandin, so that complete uterine evacuation can be produced with a low dose, thereby minimizing the systemic side-effects of the prostaglandin such as vomiting and diarrhoea. Provided that this technique does not leave a proportion of women with incomplete abortions who still require a vacuum aspiration, or even cervical dilatation and uterine curettage (a 'D and C'), it could prove to be a major breakthrough. It is becoming apparent that RU 38486 has several modes of action in addition to its abortifacient effect: it can also prevent ovulation, or hasten luteal regression, apparently because of a central inhibitory action on gonadotrophin secretion. Thus, a woman could use it for once-a-month 'menstrual regulation', regardless of whether or not she was pregnant. If we had a safe and effective medical abortifacient that could be self-administered in the woman's own home, we could leave it to her to make these critical decisions, without subjecting her to the antagonistic attitudes of some doctors, lawyers, politicians and public pressure groups who would seek to interfere in what is, after all, a very personal and private decision.

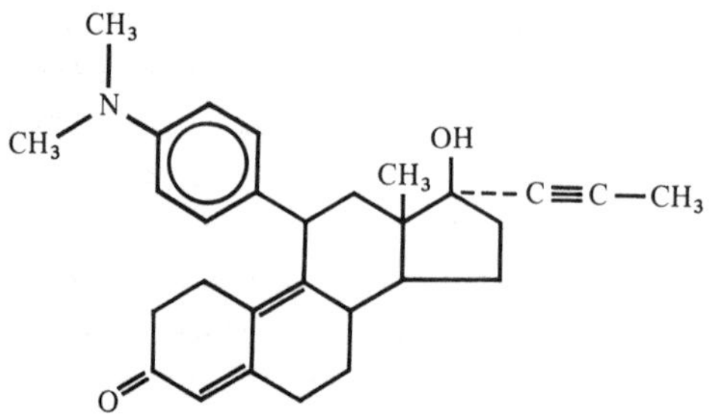

Fig. 2.21. The new steroidal anti-gestagen, RU 38486, produced by the Roussel-Uclaf pharmaceutical company in France.

Contraceptives of the more distant future?

In discussing tomorrow's contraceptives, I have tried to confine my remarks to techniques that we can be reasonably certain will become freely available in the near future. But if we are prepared to adopt a longer time-frame, there are many ideas in the air, a few of which may reach practical fruition in our lifetimes. But is that soon enough to meet the urgent needs of the developing world?

Gonadotrophin-releasing hormone analogues

We are so used to thinking in terms of steroidal contraception that it is refreshing to discover that there is another theoretical possibility – peptide contraception. When synthetic gonadotrophin-releasing hormone (GnRH) and its biologically active analogues were undergoing routine toxicology testing in animals prior to clinical use, it was noticed by Jurgen Sandow of Hoechst that in large doses these compounds had a paradoxical inhibitory effect on pituitary and gonadal activity. Although the drug companies were anxious to license these compounds for therapeutic use to stimulate ovulation in hypogonadotrophic women, and to induce testicular descent in cryptorchid children, research groups in Sweden and West Germany were able to show that daily administration of a GnRH analogue by nasal spray in a carefully metred dose to normally cycling women could interfere with pituitary function in such a way as to inhibit ovulation for the duration of treatment, in some cases for up to a year. However, there were problems from breakthrough bleeding and irregular menstruation in some women, and others remained amenorrhoeic while on treatment. This forced the investigators to think in terms of additional steroidal treatment to control the menstrual disorders, so that the technique has now become too complex and elaborate for use in a developing country.

However, there has recently been another interesting development. These same GnRH analogues have come into vogue for the suppression of testicular activity in men with prostatic cancer, and ICI have developed a biodegradable implant containing a GnRH analogue that is no bigger than a grain of rice. It is unlikely that these implants will lead to the development of a male contraceptive, since in addition to inhibiting spermatogenesis they also depress testosterone levels on which a man's libido depends; if testosterone replacement therapy is given, this will unfortunately stimulate spermatogenesis. But it is possible that these implants could provide a novel approach to female contraception, if problems of menstrual irregularity and hypooestrogenism could be overcome. Ideally, one would aim to achieve an endocrine state comparable to lactational amenorrhoea, when there was complete inhibition of ovulation, but sufficient follicular development in the ovaries to prevent the occurrence of menopausal symptoms, without causing endometrial proliferation. This may be a difficult balance to achieve.

The great advantage of peptide contraception would be that GnRH and its analogues have little if any known systemic toxicity, and are without effect on the peripheral target organs such as the breast, uterus and vascular system, in contrast to steroids.

Immunological contraception

The reproductive tract offers many potentially interesting opportunities for immunological intervention (see Book 4, Chapter 6, Second Edition). One of the most promising leads at the moment would appear to be immunization of the female against the zona pellucida which surrounds the oocytes; anti-zona antibodies have been shown to prevent fertilization following active immunization of laboratory animals and primates. The problem is to isolate and characterize a specific zona antigen, and then to use recombinant DNA or other genetic engineering techniques to produce sufficient for large-scale trials. There is also some concern that these anti-zona antibodies might have adverse long-term effects on the ovaries themselves; for example, if they increased the rate of oocyte atresia, this might bring on a premature menopause. Other unknowns include the duration of the immunity, and the feasibility of producing adequate antibody titres in all individuals who are immunized. But if all went well, zona immunization might provide us with the longed-for non-surgical female sterilization procedure.

Another promising lead is immunization against pregnancy. Human chorionic gonadotrophin (hCG) is produced by the embryo at the end of its first week of life, and is thought to represent the luteotrophic stimulus that establishes the pregnancy (see Book 3, Chapter 7, Second Edition). The hormone is composed of an α subunit which is the same as that found in FSH, LH and TSH (see Book 7, Chapter 2, First Edition), and a specific β subunit. Studies in baboons and marmoset monkeys have shown that if they are actively immunized against the β subunit of hCG, they are unable to maintain a pregnancy provided that the antibody levels remain high. The human β subunit can be made antigenic to humans by coupling it to a larger foreign protein, like diphtheria toxoid. The World Health Organization is now initiating clinical trials in the United States and Australia to see initially whether satisfactory anti-hCG antibody titres can be safely achieved in women who are not at risk of pregnancy. Only if this proves to be the case will further trials be undertaken to assess its contraceptive efficacy. There are many potential pitfalls along the way, such as undesirable cross-reactivity of the antibody with other tissues, failure to produce an adequate antibody titre, or to sustain it for a sufficient period of time, variability in response between individuals, and the possibility of life-threatening incomplete late abortions as antibody titres wane.

Finally, there is the possibility of immunizing women against the spermatozoa of their partners. It is paradoxical that we recognize anti-sperm antibodies in women as a factor that can contribute to infertility, and yet it has proved difficult to induce infertility experimentally by immunizing animals against spermatozoa. Perhaps if we could isolate and purify the appropriate sperm antigen, and cause the female reproductive tract to secrete high levels of antibody directed against it, we might be more successful.

There are two alternative philosophies concerning contraceptive research and development. On the one hand, there are those who, like the poet Geoffrey Scott, believe that:

> All we make is enough
> > Barely to see
> A bee's din,
> > A beetle-scheme –
> Sleepy stuff
> > For God to dream:
> Begin.

They would encourage us to redouble our research efforts in an attempt to discover that magical perfect contraceptive without which we cannot begin to solve the world's population problems.

And then there are those who would agree with the poet's concluding stanza:

> All we know is enough;
> > Though written wide,
> Small spider yet
> > With tangled stride
> Will soon be off
> > The page's side:
> Forget.

Perhaps we should abandon our search for something better, and be content with the contraceptives that are already available. After all, the developed countries have used them successfully to achieve near zero rates of population growth, and China is making a valiant effort to follow suit.

The solution must surely be a compromise between these two extremes. We need to improve the distribution of existing contraceptives, and we need to design better ones, with the developing countries of Asia, Africa and South America specifically in mind. The commitment and funding for this work can only come from the developed countries. Unfortunately, for them overpopulation has become yesterday's problem; they are now more concerned with the threat of nuclear war and its ensuing nuclear winter. We must therefore rekindle a sense of urgency about the population problem, which is far from solved. We cannot afford the increasing polarization that is taking place between the haves, who are becoming wealthier, and the ever-increasing numbers of the have-nots; it will tear our little world asunder. We have a people problem, here and now.

Suggested further reading

1984– and after. *People*. Vol. 11, No. 1. International Planned Parenthood Federation; London (1984).

Time bomb or myth: the population problem. R. W. McNamara. *Foreign Affairs*, **62**, 1107–31 (1984).

Seeking better contraceptives. S. J. Segal. *Populi*, **11**, 24–30 (1984).

Breast Feeding. R. V. Short. *Scientific American*, **250**, 35–41 (1984).

Breast cancer, cervical cancer, and depot medroxyprogesterone acetate. WHO Collaborative Study on Neoplasia and Steroid Contraceptives. *Lancet*, **ii**, 1207 (1984).

Fertility in Adolescence. Ed. A. S. Parkes, R. V. Short, M. Potts and M. A. Herbertson. *Journal of Biosocial Science*, Supplement 5 (1978).

Population Growth and Economic and Social Development. A. W. Clausen. The World Bank; Washington, USA (1984). (Available free on application to World Bank offices.)

China Population Newsletter. Vol. 1, No. 1 (1983).
    Published from the China Population Information Centre, available from Chinese Embassies.

Population Reports, in particular the following:
    Minipill – a limited alternative for certain women. Series A, No. 3 (1975).
    Long-acting progestins – promise and prospects. Series K, No. 2 (1983).
    Periodic abstinence: how well do new approaches work? Series I, No. 3 (1981).
    Vasectomy – safe and simple. Series D, No. 4 (1983).
    Breast-feeding, fertility and family planning. Series J, No. 24 (1981).
    Population and birth planning in the People's Republic of China. Series J, No. 25 (1982)
    IUDs: an appropriate contraceptive for many women. Series B, No. 4 (1982).
    Oral contraceptives in the 1980s. Series A, No. 6 (1982).
    New developments in vaginal contraception. Series H, No. 7 (1984).
These reports may be obtained free of charge on application to the Population Information Program, The Johns Hopkins University, Hampton House, 624 North Broadway, Baltimore, Maryland 21205, USA.

Report of the Public Board of Enquiry on Depo-Provera. J. Weisz, Chairperson, (1984).
Available on request from the US Food and Drug Administration, Washington, USA.

World Health Organization. Annual Reports of the Special Programme of Research, Development and Research Training in Human Reproduction.
These reports may be obtained free of charge on application to the World Health Organization, Publications Department, Geneva 1211, Switzerland.

The Politics of Contraception. C. Djerassi. Norton & Co.; New York and London (1979).

The State of the World's Children. J. P. Grant, UNICEF and Oxford University Press (1984).

Birth Control Technologies. M. J. K. Harper. Heinemann Medical Books; London (1983).

Textbook of Contraceptive Practice (2nd ed.). M. Potts and P. Diggory. Cambridge University Press (1983).

Abortion: Medical Progress and Social Implications. Ciba Foundation Symposium 115. Pitman; London (1985).

The Antiprogestin Steroid RU 486 and Human Fertility Control. Ed. E.-E. Baulieu and S. J. Segal. Plenum Press; New York (1985).

Handbook of Family Planning. Ed. N. Loudon. Churchill Livingstone; Edinburgh (1985).

3

Contraceptive needs of the developing world

D. M. POTTS

Homo sapiens, by the standards of other mammals, has a low reproductive potential. Yet we are the first large mammal in biological history whose reproductive achievement presents a threat to our own survival – and to many other vertebrate and complex life forms: we add a million to our numbers every 5 days and exterminate 100 other animal species each year. It is a paradox that our inefficient human reproductive system, with its slow onset of sexual maturity, frequent and prolonged intervals of infertility, and obstetric hazards, has become the source of extraordinarily rapid population growth.

Currently, there are over 80 million more births than deaths in the world annually. That this phenomenon is a recent one is underlined by the fact that contemporary population growth is so rapid that more than half the world's population are below the age when reproduction normally begins (Fig. 3.1). The broad base to the world's population pyramid is also a reason why, although the percentage rate of population change began to fall in the 1970s, the absolute annual addition to the world's population continues to rise. By the year 2000, it will peak at approximately 100 million a year – a new Pakistan, annually. Ninety-two per cent of global population increase between 1980 and 2000 A.D. will take place in developing countries. Whatever steps are taken and however rapidly fertility declines, the world population is set not only to double, but to grow to between 8.4 and 12.4 thousand million by the middle of the twenty-first century.

The scale of recent changes easily blinds us to the fact that most of human history has been lived in a world where the human parent saw no more than two children grow, mature and have children of their own. Large families *surviving* to the next generation are a unique and temporary aberration in the pattern of human reproduction.

The demographic transition in the West began with a phase of slowly declining death rates and relatively high fertility, followed by a long interval of slow fertility decline against a background of relatively low deaths, and finally by a recent phase of balanced births and deaths. Western populations grew from a smaller base, expanded more slowly and enjoyed the demographic safety valve of overseas emigration. However, until recently, family planning was often opposed by statute law and conducted without help from the medical profession.

In the contemporary developing world, death rates have fallen more rapidly than they ever did in the West and immigration across frontiers is becoming increasingly restricted. Internal migration, however, is accelerating and by the year 2000 an estimated 1500 million people – equal to the total population of the world in 1900 – will live in Third World cities. The positive factors in favour of the rapid adjustments the developing world must make is that most countries have adopted official family planning policies, and a greater range of effective methods of fertility control is available than when birth rates began to fall in the West.

How long will it take for the world population to reach equilibrium and will this come about as a result of continually falling birth rates or because of a rise in death rates?

Existing restraints on fertility

Important limitations on human fertility exist in nearly all communities and must be understood if modern contraceptive choices are to be

Fig. 3.1. Populations of less and more developed regions in 1975 and 2000 AD. (Modified from *The Global 2000 Report to the President of the US* (1980).)

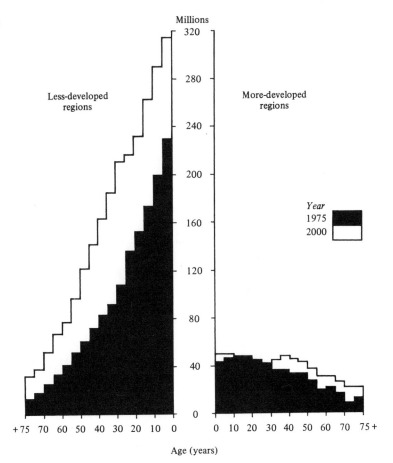

introduced effectively (Fig. 3.2). The !Kung (Book 4, Chapter 2, Second Edition), who use no birth control whatsoever, but whose way of life parallels that of 98 per cent of human biological history, have a population doubling time of 300 years. The women of the Gainj people in Papua New Guinea do not begin ovulatory cycles until, on average, age 20 and have a total fertility rate of 4.3.

Before the demographic transition began in Britain, society had undergone a profound transition from the extended to the nuclear family. In Elizabethan times, up to a third of all boys and a quarter of all girls lived outside their own family as apprentices or servants. Wrigley (see Suggested further reading) has shown that in Colyton, Devon, between 1647 and 1719, the mean age at first marriage was 28 for men and 30 for women. By contrast, in 1975–77, the median age of women at first marriage in Nepal was 15, in Pakistan 16 and in Thailand 19.

In most Muslim countries, there is little or no premarital intercourse, but in some other communities, such as the Caribbean and parts of Africa, premarital teenage intercourse is common. In the shantytowns of the exploding cities of the Third World, traditional societal values can break down and young women, whether married or unmarried, are increasingly exposed to the risk of pregnancy. The woman who marries as a teenager not only faces more childbearing years than a woman marrying later but, in human terms, finds it more difficult to control her fertility. Commonly,

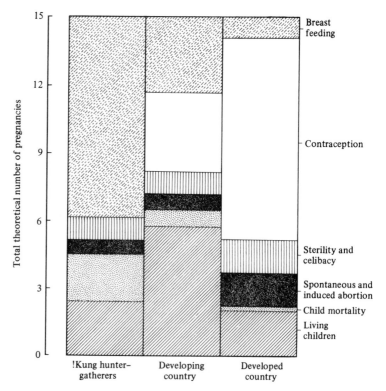

Fig. 3.2. Natural and artificial restraints on fertility. The figure assumes that, in the absence of breast feeding or any method of contraception, a woman could have 15 pregnancies.

she marries a man 10 or more years older than herself and her educational and work opportunities are usually curtailed by early childbearing.

In some countries, however, the spread of education and urbanization is pushing up the age of marriage. In Sri Lanka, one-third of the fertility decline over the last two decades has been because of the rising age of marriage. In Korea and Taiwan, changes in the median age of marriage have been more important determinants of fertility in the 15- to 24-year-old age group than adoption of modern methods of contraception. The People's Republic of China has taken a number of legal and social steps to delay marriage until at least 23 for women and 25 for men, and in some provinces more than 90 per cent of marriages fall beyond this boundary.

In many countries, lactational amenorrhoea remains the single most important restraint on fertility. Unfortunately, the inhibition of ovulation in *Homo sapiens* has proven to be the most labile aspect of human reproduction. Urbanization, perceptions of what is modern and good, aggressive advertising, hospital delivery and arbitrary and sometimes counterproductive medical advice have all encouraged the spread of bottle feeding. In mammals that breed throughout the year, lactation ensures the optimum spacing of pregnancy, either by the delay of implantation or by inhibition of ovulation, but relatively small changes in the pattern of human suckling have a large effect on gonadotrophin and prolactin levels, which in turn control ovulation. As shown in Book 3, Chapter 8, Second Edition, whether the child sleeps with his mother at night and the patterns of supplementary feeding both influence the return of ovulation during a given interval of lactation. Among the nomadic !Kung, the pregnancy interval is 44.1 months. The Koran states that 'Mothers shall give suck to their offspring for two whole years', summarizing a pattern once universal in traditional societies but now disappearing.

Modern artificial methods of contraception play two separate roles: to replace what long intervals of lactation once did in spacing pregnancies and to give a new degree of control over fertility. In some developing countries, curtailment of traditional practices of breast feeding is taking place more quickly than the adoption of new methods of contraception. Egypt spends over $20 million a year subsidizing milk formulae, which raises fertility, and less than $5 million a year per family to bring fertility down. Serious problems of high fertility will be made worse unless high levels of breast feeding can be maintained (Table 3.1).

Within a sexual partnership, and apart from intervals of anovulation associated with pregnancy and lactation, nearly all human societies know and practice certain additional restraints on fertility. Abstinence is enjoined on priests in several religions (in Tibet a generation ago, one quarter of all adult men were married monks) and celibacy is admired within marriage by some Hindu groups, where the idea of sleeping with one's wife only a few times in a lifetime (while probably excessively rare) is sometimes considered an ideal, associated with exceptionally healthy offspring.

Abstinence is commonly expected at the time of religious rituals. In parts of Africa, Indonesia and Bangladesh, men refrain from intercourse during much of their wives' lactation, just as they were exhorted to do in sixteenth- and seventeenth-century Europe. Among the West African Yoruba a generation ago, post-partum abstinence lasted 3 years, and even today the average is 27 months in the countryside. Women who break the taboo are called 'animal-like' or 'sex-crazed', and ostracized by relatives. A generation ago, the Meru of Kenya expected a couple to abstain sexually after becoming grandparents.

Coitus interruptus is known in many societies, particularly Catholic and Muslim. The Prophet Mohammed approved the use of coitus interruptus (*al-azl*) in the Koran. In Turkey in the 1960s, the percentage of couples reporting the use of coitus interruptus rose from 14.5 to 25.2, in contrast to the adoption of oral contraceptives, which rose from a minimal 1.1 per cent to only 2.3 per cent. As in Western Europe 100 years ago, coitus interruptus is often first used by the upper classes in the contemporary Third World, being common, for example, among doctors in India. The reported failure rates vary widely, from as low as 3 to as high as 17 per 100 woman-years (Fig. 3.3). The method costs nothing, cannot be left behind when a couple travel and has no known side-effects. Coitus interruptus is like a buffalo cart – no doubt there are better methods of transportation or better methods of contraception, but for a great number of people it still remains the most practical solution to an everyday problem. In Italy in the 1980s, it remains the second most common method of family planning.

Coitus reservatus, or vaginal intercourse without ejaculation, is practised in some Oriental communities, especially among older couples where loss of semen is traditionally regarded as something sapping a man's strength.

Table 3.1. Rise *in contraceptive use required if duration of breast feeding* falls, *in order to maintain* current *marital fertility rates in selected countries*

| Country | Total family size | Current situation (1975–78) | | Future use of contraception required if post-partum amenorrhoea falls to | |
		Mean duration post-partum amenorrhoea (months)	Use of contracep- tion (%)	6 months (%)	2 months (%)
Kenya	7.7	10.7	7	24	38
Pakistan	6.9	13.3	6	30	43
Bangladesh	6.4	18.5	9	43	54
Colombia	5.3	7.2	37	41	52
Costa Rica	4.0	5.3	59	—	66

Induced abortion is a universal phenomenon. No country has seen a decline in fertility without considerable recourse to abortion and no country seems likely to see its fertility fall in the next 20 to 30 years without a significant, and often increasing, number of abortions taking place.

Traditional methods of abortion divide into three groups. The desire for a medicine that will bring on a delayed period is widespread. Many of the remedies sold probably have no pharmacological effect and merely exploit the unmet demand for family limitation. However, some, such as zoapatle (*Montanoa tomentosa*) used in Mexico, have a physiological action on the uterus. Herbal abortifacients are routinely used in the family planning programme of the People's Republic of China. Most societies know that the insertion of a foreign object into the cervix will initiate abortion. The use of the sonda (or urinary catheter) is exceptionally common in Latin America, and approximately one million women are admitted to hospitals in the continent every year suffering from the effects of criminally induced abortion, mostly initiated by intrauterine manipulation. Massage abortion is a technique limited to the Orient and is known from the Philippines through to Burma and south to Indonesia. The traditional birth attendant usually treats the woman seeking the abortion in her own home, where she lies on the split bamboo floor of her house (Fig. 3.4). Patiently, and with increasing force, the midwife massages the abdomen, attempting to reach behind the pubic bone and put the maximum pressure on the uterus. She will continue until pain interrupts the procedure, or until blood appears *per vaginam*. Sometimes she will use her elbows instead of her hands, occasionally her bare feet or the instrument with which Asian villagers pound their rice. Side-effects include haematuria, bleeding from the rectum and an appendicitis-like syndrome where the woman may be admitted to the hospital with fever, abdominal tenderness and rigidity and require hysterectomy. A survey in Thailand showed that in the villages alone there are perhaps 250 000 induced abortions every year, of which 80 per cent are massage abortions.

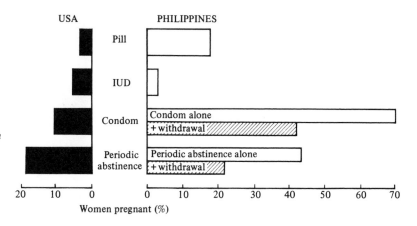

Fig. 3.3. Contraceptive failures in a developed and a developing country. Percentage of women experiencing a pregnancy in the first year of the specified method in the USA and the Philippines. (Data from J. Laing, personal communication, and *Population Information Program*, Series 1, no. 3 (1981).)

Pressure to control fertility

Urbanization, education (particularly of women), increasing *per capita* income and other markers of modernization are all demonstrable correlates of fertility decline. There is no doubt that fertility falls as society gets rich and modernizes, whether there is political commitment to family planning, as in contemporary Singapore, or antagonism, as in late nineteenth- and early twentieth-century Europe. At the microeconomic level, Caldwell (see Suggested further reading) has pointed out that in many traditional societies a child brings economic benefits to its parents from an early age: in Kenya, for example, a 6-year-old is expected to fetch firewood and water or help clean the house. As education spreads, a child turns into an economic liability, at least until the teens and perhaps longer, and this reversal of economic forces appears to stimulate fertility restraint (Fig. 3.5). However, experience in many countries shows that socio-economic development is not a prerequisite for fertility decline.

The problem facing the modern world is that population growth retards economic progress, whether capitalistic or communist, particularly for the poorest members of society. Explosive population increase requires that

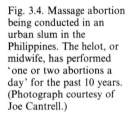

Fig. 3.4. Massage abortion being conducted in an urban slum in the Philippines. The helot, or midwife, has performed 'one or two abortions a day' for the past 10 years. (Photograph courtesy of Joe Cantrell.)

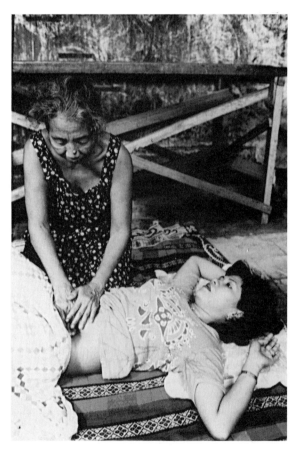

a significant proportion of fixed capital formation be reinvested to build the houses, roads, health services and educational systems necessary to accommodate the ever-increasing population. For example, the annual population growth in Nigeria (1965–70) was 2.5 per cent: capital formation in the country was 12 per cent of the gross national product (GNP), but 7.5 per cent of the GNP had to be reinvested to cope with population increase, i.e. nearly two-thirds of the capital accumulated was devoted to maintaining the economy, rather than moving it forward.

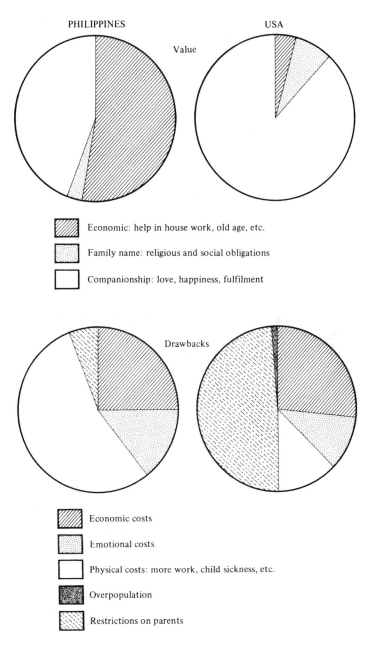

Fig. 3.5. The perceived value and drawbacks of children in the Philippines and the USA (1975). (Modified from *The World Bank Development Report, 1984*.)

The high dependency burden, low wages and a large, poorly educated population with a great deal of unemployment all make it more difficult to accumulate capital in a country with a high rather than a low population growth rate. Therefore, the gap between rich and poor, both at a global level and between individual countries, grows larger year by year (Fig. 3.6). The GNP of the USA is approximately equivalent to the combined GNP of all the developing countries and this relationship seems likely to remain over the next two decades. Because of differentials in population growth, the gap in *per capita* income between the developed and the developing world will grow even wider. Poverty is defined by the Indian government as having less than $7.50 a month, and 40 per cent of the population fails to reach this pathetic line. Within countries, comparable problems exist in the distribution of wealth (Table 3.2).

Table 3.2. *Differentials in economic growth in a developing country* (*Peru, 1950–75*)

	Percentage growth		
	GNP	Population	*Per capita* income
Rural economy	2.4	0.8*	1.6
Slum urban	4.6	4.5	0.1
Modern urban	6.6	4.4	2.5

* High population growth rates are held down by migration to urban areas.

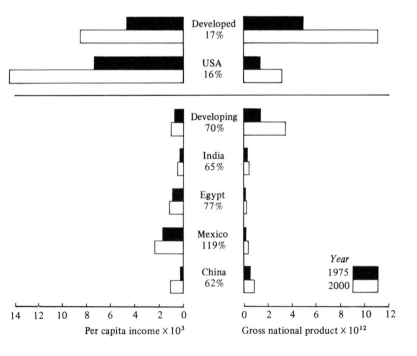

Fig. 3.6. Projected growth of *per capita* income and gross national product (1975 US constant dollars) in developed and developing countries. (Modified from *The Global 2000 Report to the President of the US* (1980).)

Population growth is not the sole cause of poverty around the world, and conversely, lower birth rates will not automatically bring wealth or equitable income; distribution, population and economic change are inextricably mixed. The need is to develop a strategy that will bring about population decline in the presence of slow or stagnant economic growth. Fortunately, the availability of the means to control fertility appears to go a considerable way to bringing about fertility decline. Family planning should be welcomed as a short-cut to certain aspects of development. In Egypt, where family planning was specifically linked to development programmes, contraceptive prevalence grew more slowly than in other countries where emphasis was placed on services.

World Fertility Survey

The World Fertility Survey (WFS) was a unique international collaborative effort conducted in the mid-1970s when 52 countries, assisted by the UNFPA and the US Agency for International Development (USAID), asked comparable questions of a national sample of households. The study focused on desired as well as achieved family size. There may have been some biases in reporting, but they were most likely towards underreporting unwanted births. Significantly, in nearly all countries, achieved family size outstrips desired family size; the desire for family planning services is already there.

One-half to two-thirds of the women in countries as diverse as Colombia, Panama, Peru, Indonesia, Korea and Sri Lanka have had one or more unwanted births by the end of their fertile life (Fig. 3.7). In practically every

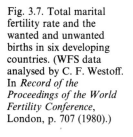

Fig. 3.7. Total marital fertility rate and the wanted and unwanted births in six developing countries. (WFS data analysed by C. F. Westoff. In *Record of the Proceedings of the World Fertility Conference*, London, p. 707 (1980).)

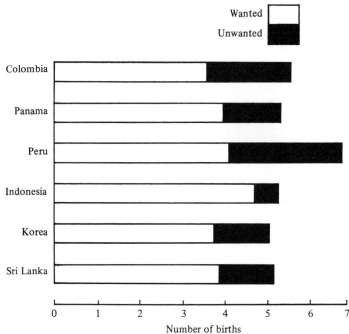

part of the developing world, there is an immediate, large, unmet need for contraceptives. The prevention of all unwanted births would reduce marital fertility in most countries by one-quarter to one-third. In Pakistan, the average woman wanted 4.3 children, but the total married fertility rate was 7.1, i.e. each woman was having, on average, 2.8 unintended or unwanted children.

Throughout the world (excluding China and the USSR), contraceptive use grew from 101 million between 1971 and 1976 to 136 million users, yet family planning programmes did not keep up with the growing world population and the estimated number of women at risk of pregnancy and *not* using contraceptives rose from 263 million in 1971 to 266 million in 1976. Globally, as many as six out of ten women aged 15–44 are at risk of unwanted pregnancy, i.e. they are sexually active, not practising contraception and neither pregnant nor trying to conceive.

Health

The need to control fertility has been elevated to the status of a basic human right. But, in addition to permitting the individual to exercise one of the most fundamental of choices, and apart from the role that fertility decline must play in economic progress, family planning also makes a remarkably powerful, cost-effective contribution to maternal and child health. In the next 20 years, it is estimated that 3000 million babies will be born into the world; 2000 million of these births will not be attended by a trained person (obstetrician or midwife). Studies by Family Health International (FHI) in areas of Indonesia and Egypt found that 20–28 per cent of all deaths to women aged 15–44 are pregnancy-related. The mortality at an unattended delivery in the developing world is probably somewhere between 1 in 100 and 1 in 500, which means that between two and ten million women will die in a holocaust of maternal deaths in the next 20 years.

Women of higher parity and in the older age groups are often those who adopt contraception first, and these are also the groups most at risk from death in childbirth (Fig. 3.8). The same pattern is repeated in infant mortality. Short pregnancy intervals (2 years or less) are also associated with higher maternal and infant deaths. In Egypt, a child born 2 years or less after its sibling is more than twice as likely to die in the first 5 years of life than one born 4 or more years later. The larger the family size, the greater the risk of gastroenteritis and other causes of child loss. Therefore, family planning is an essential part of primary health care.

In contrast to someone with a disease, the consumer of family planning therapies makes his or her own diagnosis: 'I believe I am fertile and I do not want a pregnancy.' Contraceptive methods are relatively simple; fatal accidental poisoning is impossible and the dose of oral contraceptives does not have to be adjusted for the individual, as, for example, insulin does. Therefore, family planning can often be the first element of primary health care to enter a traditional society and should be welcome for that reason.

Resources for family planning

Financial

Currently, approximately $1000 million a year is spent on family planning in the developing world (excluding China). In a country such as Korea, where fertility has declined and *per capita* income increased greatly, the national government provides most of the family planning funds ($6 million 1975–76, or 22 cents US *per capita*), while in a poor country such as Pakistan, external funding commonly exceeds government appropriations (1975–77: government $8 million, outside sources $16 million). In Indonesia in 1980, $84.4 million was spent on family planning, of which $34.7 million came from external donors. In the late 1970s in Guangdong Province in China (population 55 million), approximately $6.6 million, or

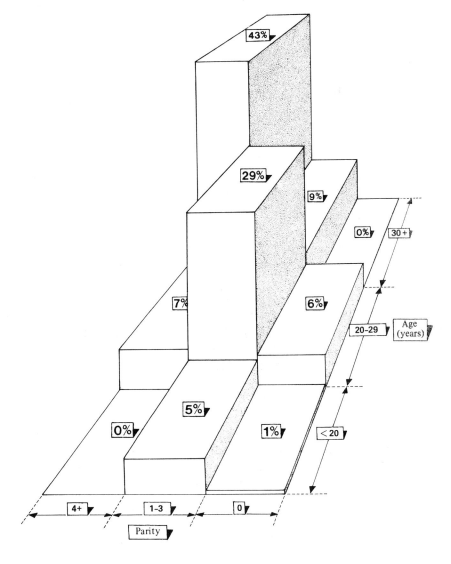

Fig. 3.8. Maternal deaths by age and parity. Menoufia Governorate, Egypt. (Family Health International data, 1982–84.)

12 cents *per capita* annually, was put into family planning. For reasons that will be looked at later, expenditures do not correlate with achievements. In the sample years 1975–77, Thailand spent 9 cents *per capita* and Pakistan 34 cents *per capita* on family planning, but the progress in the two countries was the reverse of the investment.

Personnel

Western-style physicians in developing countries represent a heavy capital investment and expect a high standard of living; therefore, most of them practise in urban areas, often predominately in the capital city. Others emigrate to the West (18 per cent of new graduates from India, 67 per cent from Thailand): there are as many Indian doctors practising in the National Health Service in Britain as there are for 500 million people in rural India.

In some cases, shortages of trained midwives and auxiliary personnel are even more serious than shortages of medical practitioners. In Iran, before the revolution, there were 11 000 practising doctors but less than 700 fully trained midwives. In 1966, the government of India decided to institute a core of 180 general duty, family planning women officers, but by 1971 only 29 had been appointed. Any method of family planning that is dependent upon sophisticated medical skills will be geographically and economically inaccessible to the rural populations and may not even penetrate into the urban slums and shanty towns.

Fortunately, poor communities in the developing world have their own pattern of medical care that could be appropriate for the dissemination of modern methods of family planning. In the Indian subcontinent, traditional practitioners (ayurvedic and homeopathic) are found in most villages, have a formal training and often deal in medicines related to fertility, impotence and abortion. Traditional midwives can be taught the elements of safe obstetrics and play a role in family planning. Unfortunately, these options are not always exploited.

Shops are certain to remain the largest immediately available outlet for family planning for the foreseeable future. There is a higher ratio of retailers to the rest of the population in a poor country than in a rich one. Every little community has its store (Fig. 3.9). The commercial trade in contraceptives is important in many countries, but prices are often beyond the economic reach of the poor. The social marketing of contraceptives, where the price is subsidized centrally, has proved remarkably successful (Fig. 3.10). Nationwide coverage can be established rapidly, advertising skills and wholesaling facilities already exist, and social marketing, while making contraceptives widely available in the culturally acceptable way, can avoid the political and religious opposition sometimes aroused by more expensive, structured programmes imposed from outside. In Bangladesh, a project initiated by Population Services International (PSI) now sells 5 million condoms monthly.

Fig. 3.9. The local retailer is often an appropriate person to distribute contraceptives: here, Pills are being sold in one of the many floating markets in Thailand. (By courtesy of *People*.)

Fig. 3.10. Social marketing of contraceptives in Bangladesh. Couple-years of protection (1976–82). (Data from *Population Services International*.)

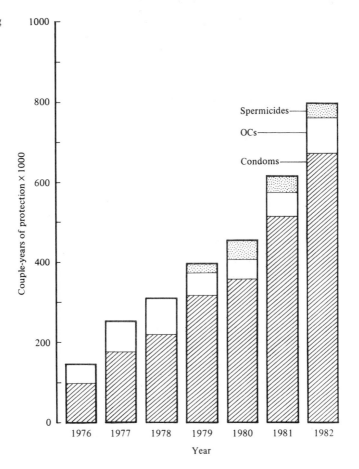

Local production of condoms and oral contraceptive tableting, although not total manufacture, take place in some developing countries. Only Mexico and the People's Republic of China undertake the full synthesis of steroids.

Management

The government infrastructure in many developing countries is weak. There is usually a greater social and political distance between the government administrator and the peasant, or slum dweller, in a developing country than there is between the same person and the citizenry of the developed world. Commitment to the job is sometimes weak. A civil service appointment often represents security and a pension for the individual, and many national family planning programmes have failed because the government appointees prefer to sit in their offices rather than deal with problems that can be solved only by going into the community. For the few people with a high degree of managerial skill, or considerable technical skill, such as physicians, government salaries are invariably low in relation to the rewards that can be obtained in the private sector, in the UN system or by emigration.

In the case of curative medicine, services continue to be used despite serious weaknesses. Pain and fear will drive a sick person to endure long waits and a brusque response from medical staff in government centres. In the case of family planning, those who are most vulnerable to unwanted fertility are often the very individuals least likely to travel any great distance to a family planning clinic and most likely to give up waiting in line if the doctor comes late, or leaves early, to attend to private practice.

Shortcomings in management systems are unlikely to be overcome in the next one or two decades, particularly in countries of greatest need. Therefore, alternative ways of making contraceptives available, such as through social marketing schemes and by coupons to reimburse private doctors for IUD insertions or voluntary sterilization, need to be used. A family planning association or institution, such as the Community-based Family Planning Services in Thailand, can offer a cost-effective structure, encourage competent management, fire staff who do not perform and, when they offer a needed service, win the love and respect of the community.

Contraceptive methods

In the West, family planning services developed following World War I, more than a generation after fertility began to decline, yet societal hostility was so strong until the 1950s and 1960s that services were unable to offer sterilization and stood aside from the problem of abortion. When population growth took off in the developing world, the Western world had still not arrived at any balanced and open attitudes toward contraception. Not surprisingly, the expert advice that was given to developing countries in the 1960s and 1970s often turned out to be inappropriate. Conventional methods of contraception, such as the condom, were 'undersold', the medical input into oral contraceptives was overemphasized and the limitations of intra-uterine devices in a traditional society were misunderstood. Above all, there was a lack of realism about the role of sterilization and abortion in fertility regulation.

Reversible contraceptives. On the whole, the new methods of contraception that Western medicine was able to offer have been used unimaginatively. Hindu women are not supposed to cook their husbands' food while menstruating and Muslim women cannot enter the mosque, but little effort has been made to use the ability of the Pill to control patterns of menstruation, *as well as* inhibit ovulation. Contraceptives were given to the individual who was medically involved rather than to the social decision-maker; yet, in Japan, 20 per cent of the condoms sold are sold by women to women, while in Bangladesh, most contraceptive pills are sold by men to men, to pass on to their wives. The Western press and medical profession have played up the side-effects of oral contraceptives, confusing the public and virtually destroying the use of oral contraceptives in several countries, such as the Philippines.

The risks and benefits of contraceptive use, and particularly those of oral contraceptives, are discussed in Chapter 4. On the evidence currently available, the protective action of the Pill against ovarian and endometrial cancer appears to offset the risk due to cardiovascular complications in women under 35 and among older women who do not smoke. In the developing world, a number of new factors apply: maternal mortality is much higher (Table 3.3), and abortion remains illegal and, therefore, dangerous in many countries; fewer women smoke than in the West, but the facts that contraceptive distribution is in less well-trained hands and emergency medical services to deal with possible rare complications are less available weigh in the opposite direction. Overall, though, oral and other contraceptives appear to increase the average expectation of life among nearly all women in all age groups (Fig. 3.11).

Whereas many Westerners expect to receive their medicines in the form of tablets, most people of the Third World have a deeply ingrained faith in anything given from a syringe. Injectable contraceptives can cause serious menstrual irregularities and are associated with a slow return of fertility, but when they have been made available, they have proved exceptionally popular. Perhaps the most immediate step that can be taken to meet some of the unmet needs of family planning in the Third World would be to distribute an appropriate injectable contraceptive through government and social marketing programmes. Every little drugstore and injection doctor in the Third World currently giving penicillin injections would welcome such an addition.

Although the injectable contraceptive depo-medroxyprogesterone acetate (DMPA) – trade name, Depo-Provera – has had more than 11 million users around the world (half in developed countries), has performed somewhat better than the Pill at a similar stage of duration and magnitude of use, and is accepted in many European countries and in New Zealand, US Food and Drug administration (FDA) non–approval has made the drug controversial. Use was recently stopped in Zimbabwe, just as DMPA reached unprecedented levels of acceptability.

Fertility awareness methods involve a knowledge of ovulation derived

Table 3.3. *Maternal mortality in selected countries. Maternal mortality is defined as including direct causes (e.g haemorrhage) and indirect causes (e.g. a woman who died of heart disease during pregnancy)*

Country	Rate per 100 000 live births	Rate per 100 000 women aged 15–44
Switzerland	6.8	1.0
England and Wales	9.0	1.7
USA	9.6	1.9
Thailand	NA	23.9
Guatemala	90.9	NA
Egypt	82.3	NA
Egypt*	190.0	30.8

NA = not available.
* Official statistics for developing nations usually under-report deaths, and a community survey for one area of Egypt (1981–83) gives higher rates.
(*Sources:* United Nations, *Demographic Yearbook* (1983), and World Health Organization, *World Health Statistics* (1983).)

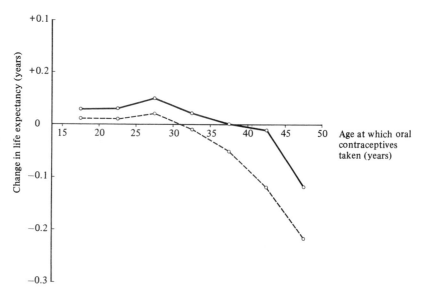

Fig. 3.11. Lifetime risks and benefits of oral contraceptive use. Change in life expectancy attributed to 5 years' oral contraceptive use at ages specified. Note that most Pill users are in the younger age groups where the benefits of oral contraceptive use outweigh risks. (USA, 1978 o–––––o; Egypt, 1979 o――――o.) (Dr J. Fortney, personal communication.)

from daily observation of cervical mucus (the Billings methods) and/or body temperature measurements. They can be used to promote fertility or, combined with periodic abstinence, condoms, spermicides or coitus interruptus, to restrict it. Like other methods, fertility awareness methods have suffered from undue controversy and are still finding their place in national programmes. In countries such as the Philippines and Sri Lanka, up to half those using any method of fertility regulation claim they are following some sort of periodic abstinence.

Voluntary sterilization. By 1978, 260 million voluntary sterilizations had been performed world-wide and the demand for sterilization continues to rise.

Current techniques of vasectomy were refined in India. As the vas deferens is placed immediately below the skin of the scrotum, the operation can be done under a local anaesthetic with minimal surgical facilities and, if necessary, by trained auxiliaries.

Female sterilization involves opening the abdomen, and the requirements for anaesthesia, and for surgical skills to perform a laparotomy and facilities to deal with emergencies, all make it more difficult to meet the demand for female than male sterilization. A transcervical method of occluding the Fallopian tubes might be the single most significant improvement in the cafeteria of family planning methods that could be offered to the world in the current decade. Medical attitudes to sterilization often remain confused. For example, in Brazil, doctors have been reluctant to offer tubal ligation as a mature choice to women but have been eager to perform it after Caesarean section, with the result that in some hospitals more than 90 per cent of all private patients have *every* delivery by Caesarean section. By spending money and risking an operation, a woman 'buys' herself a desired sterilization, relieves herself of any religious scruples and meets the perception of medical practitioners who offer sterilization as a treatment of a disease ('over-fertility') but not to implement an informed adult choice. The foremost policy and practical issues facing family planning programmes around the world is how to meet the need for abortion and surgical sterilization.

Abortion. Induced abortion appears to be an inescapable element in the transition from high to low fertility (Fig. 3.12). The most common pattern seems to be that of an initial rise in contraceptive use *and* induced abortion, but ending with a high use of contraception and a low, but irreducible, minimum of abortions. Rising rates of abortion are found where fertility is high but beginning to fall, as in Bangladesh, maximal rates in societies half-way through the demographic transition, as in Cuba, and relatively low rates (especially among married women) in Britain and the USA. It is interesting that in Tunisia, where abortion is legal, analysis suggests that the total number of embryos being destroyed may be *less* now that abortion is legal than it was when it was illegal.

The vacuum aspiration technique of early abortion was developed in China. It has proven pre-eminently the most straightforward and safest way of performing an abortion and is a robust, well-tested technology easy to apply in developing countries. 'Menstrual regulation', which is uterine evacuation with a hand-held syringe and a flexible plastic cannula that can be passed through the cervix without anesthesia, has proved an outstandingly simple answer to the large demand for a way to terminate an *early* (6–8 weeks since the last menstrual period), unplanned pregnancy. Ethical issues associated with abortion and menstrual regulation will be discussed later.

Research. Investment in contraceptive research is declining, as manufacturers are being driven out of contraceptive development by medico-legal hazards and a blizzard of bureaucratic regulations. In addition, public and philanthropic support is waning. Clearly, society must regulate drugs, and caution should be a watchword, but regulations must be realistic: pre-marketing requirements for introducing new drugs and devices should be lowered while post-marketing surveillance could be improved.

Policies and programmes
Of 114 developing countries, over 80 have enunciated population policies. These vary from a strong commitment to reduced population, as in India and China, to a reluctant acceptance, as in Kenya. Sometimes the changes in policy have been dramatic: until 1956, Thailand offered bonuses for

Fig. 3.12. Abortion and contraceptive practice during the demographic transition.

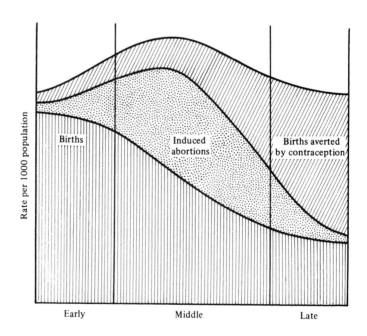

large families, whereas today there is a multimillion dollar investment in family planning programmes.

The understanding of what a population policy and its implementation means varies widely. Often, heads of state, cabinet members, senior administrators, foreign diplomats, and even advisors in international agencies, accept a statement of a population policy without inquiring about the details of the programme. Sometimes it means a commitment to make available a selected range of the reversible methods of contraception, as in the Philippines and Pakistan. Sometimes, as in India, there is an availability of contraception, abortion and sterilization, but serious limitation in implementation, largely because of an almost exclusive use of government infrastructure to implement the programme and an inability to escape from policies set by Western-trained doctors.

Family planning policies have been divided over whether there should be a vertical programme, where special staff and outreach workers are employed, or whether family planning should be integrated with maternal and child health (MCH). Unhappily, this perceived choice has led to a great deal of conflict and misunderstanding. It is logical and necessary that family planning should be part of MCH services, but these cannot provide the full range of choices necessary, e.g. to men, or to women prior to the first pregnancy. When family planning is added to busy MCH services, it tends to be pushed aside by the needs of curative medicine. Ambitious integrated health and family planning projects, such as the Danfa Project in Ghana and the Bohol Project in the Philippines, have not been attended by outstanding demonstrable success. Efforts to integrate family planning with maternal services in Kenya have been costly. A simple division of total programme costs ($36 million) by target numbers of acceptors (640000) in Kenya over a 5-year interval gives a figure of $56 for each acceptor to be recruited and an average cost of $240 for averting a birth. The *per capita* income of Kenya is under $200 and there seems to be something intrinsically wrong with a system when the cost of averting a birth equals or exceeds what many individuals earn in a year. By contrast, the social marketing project in Bangladesh costs $1.66 per couple per year of protection (plus $2 or less for commodity costs).

Family planning does not pre-empt or compete with other essential aspects of health care, but its simplicity and effectiveness make it an attractive starting point. It should not be limited to MCH services, especially when those services care for only a small proportion of the population. It can and should be the leading element in health care. Contraceptives can often be sold in the village store long before there are enough resources to build and staff a health clinic. Rarely, if ever, in the developing world should it be necessary to have freestanding family planning clinics staffed by physicians and nurses. Their skills should be held back to deal with the need for obstetric care and curative medicine. Sterilization and abortion are usually provided in secondary and tertiary health centres.

At various times in the twentieth century, Western nations have adopted pronatalist policies. Canada, Britain, France, many Eastern European countries and Russia pay extra sums of money to support children. However, these financial inducements have rarely had a significant impact upon the private behaviour of individuals. Are antinatalist policies built on financial punishment likely to work? In 1970, Singapore, as a first step in a series of penalties intended to restrict fertility, imposed charges for high-parity deliveries in hospital. In 1974, tax relief was limited to the first three children and the total fertility rate fell from 4.9 in 1966 to 2.0 in the 1980s. However, in the early 1970s, abortion was used with great caution (1969: 3.6 abortions per 1000 women aged 15–44). Sterilization was introduced in 1972, but restricted to women with a specified number of children. Full availability of voluntary sterilization and abortion was achieved only *after* the social penalties on childbearing had been instituted (1975: 23.2 abortions per 1000 women aged 15–44). It seems quite possible that the observed fertility decline might have occurred, as the means to control fertility became more realistically available, without any antinatalist policies.

Successes

China

The world is fortunate that its most populous nation has seen a dramatic decline in fertility in the last decade. In 1949, China's population stood at over 500 million and the birth rate in the upper thirties. In 1963, fertility peaked at 43 and 44 per 1000, but by 1979 the birth rate was 17.9 per 1000 nation-wide and the growth rate slightly over 1 per cent. In Shanghai (1977–78), the birth rate was 7.4, and in the Hunan Province, 15.3 per 1000. The total fertility rate (the average number of children born in a fertile lifetime) is rapidly approaching biological replacement (Fig. 3.13).

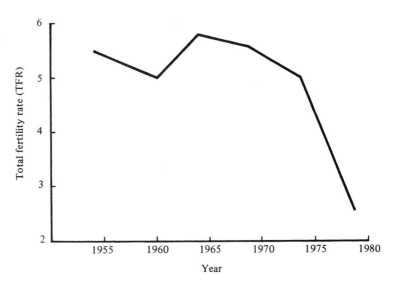

Fig. 3.13. The total fertility rate (TFR) of China during the period 1953–78. (*Population Reports*, Series J, No. 25, (1982).)

Successful as the Chinese programme has become, it started too late. Ideological opposition to family planning survived into the 1970s and a generation that needed family planning didn't get it. As a result, there are 10 million more marriages each year than partnerships are broken by death or the woman reaching the menopause. If the population is even to approach stabilization, the current generation of parents must achieve *less* than biological replacement. It is a policy that no one wants but to which it is too late to offer an alternative. To stabilize the population at 1200 million by the year 2000, 50 per cent of rural couples and 80 per cent of urban couples must have no more than one child! The one-child certificate provides for a 5–8 per cent increase in family wages (until the one child reaches the age of 14), preferential treatment in housing and food ration, and extra retirement benefits. If the couple break their pledge, they must *return* the benefits they have already received. Great social pressure is brought on any couple that defies the new social norms and, in a nation of a thousand million people, cases of frank coercion almost certainly occur. Female infanticide has always occurred in rural areas and the Chinese government has deliberately reported new cases in an effort to suppress the practice.

To outsiders, the Chinese system seems strange and threatening, but every country that fails to foresee the momentum of population growth must eventually face the same insistent demographic arithmetic – population growth is not something that can be turned off when a country is 'full', as most African leaders believe, or that responds to half measures, as Pakistan and Egypt have believed. Forestalling action must be taken at least a generation ahead. The human tragedy of China is that women who reluctantly abort a second pregnancy today are the daughters of women who were not able to gain effective access to voluntary methods of family planning a generation ago.

As in the USA and Britain, there is a wide availability of IUDs, hormonal contraceptives (where perhaps inventiveness has benefited from not having a Food and Drug Administration (FDA)), voluntary male and female sterilization, and safe abortion available through basic health services. The state provides 1–2 weeks' leave for IUD insertion, induced abortion and sterilization, and there has been a strong push to raise the age of marriage. However, while these incentives are probably useful, the availability of the full range of fertility-regulating services remains the key to success in China, and an equally impressive decline in fertility is taking place in other areas of the developing world with very different cultural backgrounds.

Indonesia

Government-supported family planning began in Indonesia in 1968. By 1976, $15 million was being spent each year and the percentage of married women using contraceptives had risen from 2.2 per cent in 1971–72 to 20.8

per cent. By 1980, approximately 35 per cent of married women were using contraceptives and family size was falling. The birth rate was 46 per 1000 in 1970 and 33 per 1000 in 1979.

The Balinese are a minority Hindu population of 2.3 million people, surrounded by 140 million Indonesian Muslims, and might be expected to be resistant to family planning. Yet between 1970 and 1976, the birth rate fell from over 40 to 27 per 1000, with more than half the women of married age, not currently pregnant, using some form of contraception.

The important thing about Indonesia is that contraceptives have been adopted in a traditional society. Improvements in education, an increase in *per capita* income and amelioration in the status of women, which are known to have been strong correlates of fertility decline in the Western world, have been modest. Of course, they are needed, but at least the country can be grateful that the family planning programme is working, ensuring that whatever economic development inputs are made will reach as far as possible. A testimony to the effectiveness of the programme, which incidently has enjoyed powerful management and the same leader, Dr Haryono Suyono, throughout, is that those with the lowest income have a higher use of contraception than both those with middle- or high-range incomes (Table 3.4).

The problem now facing the country is how to tackle the problem of illegal abortions, presently done by massage or cervical manipulation, and how to add voluntary sterilization, a much-needed option which meets with some opposition from conservative religious leaders.

Thailand

Thailand is a Buddhist country, although there are some Muslim areas in the south. Family service programmes have been even more successful than managers predicted. By 1977, just under half a million women began taking the Pill for the first time and over 100000 had a tubal ligation, 58000

Table 3.4. *Adoption of contraception in a poor society* (*Indonesia, 1977*)

Monthly income in US dollars	Percentage of non-pregnant women using a modern method of contraception
< 32	37.9
33–48	32.1
49–64	27.2
65–80	24.4
81–120	26.3
> 120	31.9

adopted injectable contraceptives and 75000 had IUDs inserted. By 1980, contraception had reached 48.5 per cent of currently married women between 15 and 44 years of age (47.2 per cent rural; 53.9 per cent urban) and the fertility rate fell from 6.1 in 1968 to 3.4 in 1978 (Fig. 3.14). In the two northern provinces, around the provincial capital of Chiang Mai, family size has fallen from 6.5 to 2.8 children over the past 20 years. This decline is partly attributed to a forceful family planning programme centred on a single hospital. Sterilization has been an accepted part of the Thai programme and there is discussion of reforming the abortion law.

South Korea

There has been a steep decline in fertility since the end of the Korean War and a revolution in income levels and modernization within the country (Fig. 3.14). Family planning began with a heavy emphasis on IUDs, and between 1964 and 1969, 55 per cent of women in the 35–39 age bracket

Fig. 3.14. Factors contributing to the reduction in total fertility rate in Thailand and South Korea. (Modified from *World Bank Development Report, 1984.*)

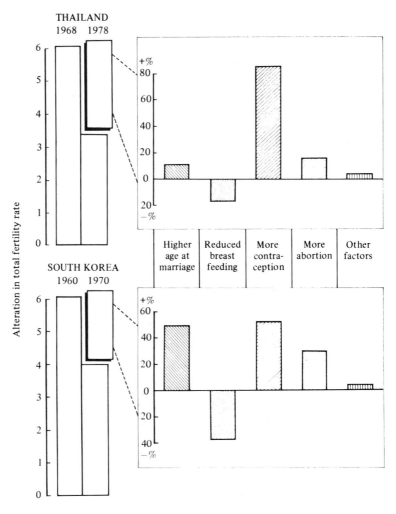

had an IUD inserted at some time. (One of the reasons for acceptance in South Korea was that doctors used vacuum aspiration abortion to back up contraceptive failures.) The Pill was added in 1968, the abortion law has been liberalized and there is a common-sense use of private practitioners who, as in Taiwan – another Asian success story for family planning – are paid on an item-of-service basis for IUD insertion and sterilizations.

Cuba

Cuba was among the first Latin American countries to begin the transition to low fertility and by the time of the 1958 revolution, the birth rate was already 26.1 per 1000. Contraceptives were made available through hospitals and polyclinics in the 1960s and use expanded with United Nations Fund for Population Activities' (UNFPA) help in the 1970s. IUDs are widely used and, unlike the rest of Latin America, safe abortion is readily available. In 1978 there were 148 220 births and 110 443 legal abortions and the birth rate stood at 15.3 per 1000. It is exceptionally unlikely that other Latin American countries will reach such a low birth rate without passing through a similar phase of high abortions.

Failures

Pakistan

Pakistan is a poor country, conservative in religion, where women are cast in a traditional role and have a low literacy rate. Government-supported family planning began in the mid-1960s but, despite several different approaches and considerable investment, the birth rate has remained stubbornly over 40 per 1000.

Undoubtedly, foreign 'experts' gave bad advice. The programme began by setting unrealistic targets based on demography, rather than by trying to maximize society's own traditions of fertility regulation. An unrealistic attempt to control fertility with IUDs was put together between 1965 and 1969. Government staff, rather than commercial enterprise or non-governmental institutions, were used. A contraceptive inundation programme in the mid-1970s planned to distribute Pills and condoms through 50 000 outlets. The commodities entered the country but many never even reached the designated outlets. As noted above, a better-managed, more commercially motivated scheme in Bangladesh has succeeded: it is not the customers but the system that failed in Pakistan.

India

India has more people than the USA, USSR and Japan put together. The one state of Uttar Pradesh has a population of 90 million and is larger than any Western European country. India continues to grow by one million additional people each month and has an annual excess of births over deaths almost equivalent to the total number of people presently living in Australia.

The logistic problems of family planning in India are enormous, with the population divided among half a million villages, many remote from any modern form of communication and totally beyond modern patterns of health care. A national family planning programme was begun in 1951, shortly after Independence, with a misguided effort to promote the rhythm method, followed by an unrealistic effort (as in Pakistan) to solve every problem with IUDs. There has been a partially successful social marketing of condoms (*Nirodh*), but no realistic use of Pills. The abortion law was liberalized in 1972, but the operation is only performed by certain trained doctors, and legalization has hardly dented the illegal abortion rate in the villages.

In 1974, Indian representatives at the World Population Conference in Bucharest launched the facile slogan, 'Development is the best contraceptive.' Within little over a year Mrs Gandhi urged her 22 Chief Ministers of the Union 'to take all necessary steps' to bring down the birth rate, concentrating on sterilization. The vote-gathering machine was used to recruit men for sterilization and the government services were ferociously driven forward to provide high numbers of operations. Seven million sterilizations were performed during the 'emergency'. Without doubt, there were gross abuses of human freedom. At the same time, even at the height of the campaign, there was probably an unmet need for female sterilization in some areas, and still no effort to market oral contraceptives or offer injectables. The excesses of the emergency rule led to the fall of the Congress Party after dominating Indian politics for 30 years. This was a dramatic example of a general failing within India of administrators in Delhi setting policies that show little respect for individuals in the countryside, particularly lower castes. The programme has always attempted to use a weak governmental structure, and persistently offered 'one shot' solutions – IUDs in the 1960s, vasectomies in the 1970s – and the overwhelming majority of Indians have never been offered a realistic, mature set of family planning choices. Yet there are many experiences to suggest that the public, which had the good sense to reject compulsory birth control heartily, would have readily accepted genuine choices in fertility regulation had they been offered. One of the success stories that could yet come from one area of south India is the adoption of the Billings method and periodic abstinence; they seemed very keen.

Issues

For a century, couples were systematically deprived by statute law and medical precedent of contraceptive options that they might otherwise have used. Margaret Sanger was arrested as a criminal when she opened the first birth-control clinic in America in 1916, and Marie Stopes was condemned from every quarter when she did the same in Britain in 1920. Developments in IUDs and steroidal contraception were delayed one or two generations after they became technically achievable. As a result, the demographic transition in the West was associated with a great deal of otherwise avoidable personal suffering. Those who fought against unequal odds to make services available were a forceful group of individualists ranging from early feminists to leftover fascists, with a solid body of wise and kind people in between.

Not only did Western families pay a great price, but the colonial powers administering most of Asia and Africa failed to introduce any aspect of birth control to balance the dissemination of death control. India's problems, for example, would have been less overwhelming if she had not had to wait to become an independent nation before tackling the difficult cultural and logistical task associated with family planning.

Happily, attitudes in the developing world are changing faster than they did in the West at a similar stage of the demographic transition. For example, in the Philippines, the importation of contraceptives was illegal until the late 1960s, yet a government clinic-based programme evolved rapidly; and, by 1975, male and female sterilization had become choices – albeit somewhat veiled ones – within the programme. It seems as if the later a country adopts a national policy, the more rapidly it moves (Fig. 3.15). Mexico, which only established government-supported family planning policies in the late 1970s, has a social marketing programme of Pills and condoms and an explosion in the use of voluntary sterilization.

Fig. 3.15. Growth in contraceptive prevalence for selected countries. The date given with each country indicates the beginning of a national programme. Every country that has promoted a service system, except Egypt and Kenya which developed integrated development/family planning programmes, has seen a rapid climb in use. (World Bank and other sources.)

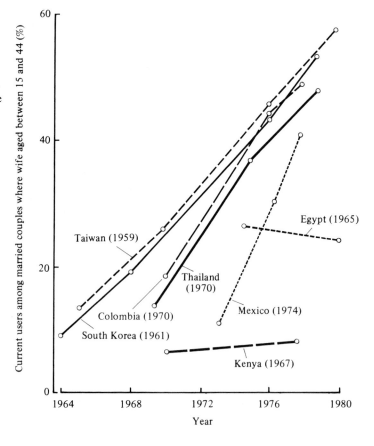

Death rates fall rapidly in response to modest monetary inputs to health. Mass vaccination campaigns and improvement in water supplies can, to some extent, be imposed on a community that may not understand their significance. Family planning, by contrast, depends on multiple, repetitive acts requiring a minimal level of community knowledge and commitment. Social, economic and educational changes and such factors as work opportunities for women are all correlated with declines in fertility, but the rate-limiting factor in fertility reduction in a great many countries – from teenagers in America to matrons in Egypt – continues to be lack of realistic access to the means to control fertility.

The WFS showed in Latin American and Asian developing countries that the mean age of the woman at the time of her last *wanted* birth was 26 to 29. It must be recognized that the reversible methods of contraception currently in use (with the possible exception of a continuous faultless use of the Pill for one to two decades) are incapable of controlling human fertility for the long interval of fertile life that exists between the last wanted birth and the menopause. In Latin America, family size was achieved within 7–8 years of marriage. Couples must either be able to choose sterilization, or have the option of abortion to back-up failures with reversible methods, or both.

If abortion and/or sterilization are available, then the available methods of contraception (Chapter 2) are adequate to delay and space pregnancies and meet any desired family size. If neither option is available, then the promotion of the reversible methods of contraception, while of value, will be associated with frustration and disappointment among many couples and with a significantly higher achieved fertility than intended. Costs will be greater.

There is a great unmet need for family planning, and a world epidemic of unwanted births that, if it persists, will send demographic shock waves into the twenty-first century. Unfortunately, family planning services often begin with the less effective methods of contraception distributed through the least accessible channels of distribution.

Family planning decision making must be placed as close to the community as possible. An exciting discovery of the 1970s was not only the community's willingness to become involved in the solution of its problems, but its commitment and ingenuity in helping to solve the problems. Lack of education is not a limiting factor when recruiting community workers, and traditional health workers and shopkeepers are the largest potential sources of manpower. Successful family planning programmes always have businesslike management, clear leadership and a rational delegation of duties. Experience has shown that it is practical and responsible to remove the reversible methods of contraception from the direct supervision of physicians, even in developed countries. A landmark policy in this field was set by the International Planned Parenthood Federation (IPPF) in 1973, which led to the validation and

expansion of community-based family planning programmes in a number
of countries.

The amount of money committed by developed countries to assist in
family planning in developing countries rose slowly, in absolute dollars,
in the 1970s, but there is a serious risk that it may decline during the 1980s.
The world spends a million dollars a minute on defence, but rich countries
donate less than one dollar a year for every woman aged 15–44 at risk of
pregnancy in the developing world.

Ethics

In the final analysis, ethics and perceptions of ethics are more important
than technology in determining the way society handles fertility regulation,
and one of the most serious weaknesses of those professionally involved
in family planning services has been a lack of explicit and informed
discussion of ethical issues.

First, modern contraception is often resisted because it is seen as
threatening changes in traditional patterns of sexual behaviour. However,
contrary to our intuition, the factors motivating contraceptive use and
premarital and extra-marital sexual activity do *not* appear to be linked.
China is a uniquely chaste society with late marriage, no prostitution, and
very little premarital sexual activity. It has a nineteenth-century horror of
masturbation. Yet all methods of contraception are as freely available as
in the USA, with its alternative life-styles, blatant exploitation of sexuality
and epidemic of premarital pregnancy. Indeed, as anyone who has worked
with Western teenagers knows, family planning services are not a Pied
Piper leading adolescents into one another's beds, but an often exhausted
runner trying to chase after and prevent damage from a life-style adopted
for other reasons, such as affluence, lack of parental control and media
example.

The second great debate over contraception revolves around the licitness
and illicitness of specific methods. Muslims condemn sterilization and the
Vatican all artificial methods of contraception, abortion and sterilization.
For the first decades of this century nearly all Christian denominations held
to Augustine's teaching that sexual intercourse was only moral for
purposes of procreation. But the Protestants in the 1920s and finally the
Catholics at the Vatican Council in 1962 accepted the expression of love
as an equal purpose of sexual intercourse, and the possibility of intercourse
at times of known infertility developed from a grudging acceptance to a
necessary part of responsible parenthood. Nearly all moral theologians
rejoice in this evolution, but the Roman leadership continues to preach
the sinfulness of 'condomistic intercourse'.

For most people, however, the battle is over and the fighting should
cease. Those who accept artificial methods of contraception should stop
poking fun at the growing movement of natural family planning and those

in that movement need to drop their antagonism to their older cousins; each has something to learn and something to give.

Thirdly, abortion raises complex ethical problems. The facts of embryology are not in dispute, but the definition of 'when life begins' is not a question open to scientific analysis. An understanding of the increasing complexity of the conceptus during gestation and of the necessary healing role that spontaneous abortion plays in mammalian reproduction, frequently eliminating abnormalities of development, may help to establish a reasoned ethical framework; but ultimately, attitudes toward abortion turn on an individual's personal faith. In any country that separates Church and State, the lawmakers can only create a framework in which individuals may follow their own religiously or privately motivated choices whether to accept or reject abortion. Certainly there is no society where effective contraception has been available, where women have consistently chosen abortion rather than contraception.

Family planning is about choices. It cannot be imposed from above, like putting in a sewer line or vaccinating pre-school children. The most successful programmes entail considerable community involvement – whether through the ancient Banjar system in Bali, the Zilla Parishads in Maharashtra, India, the Concerned Women organization in Dhaka, Bangladesh, or the village shopkeepers and distributors in Thailand, Brazil or Colombia. Socially, morally and technically, the only people available to give genuine help to the poor are usually the poor themselves.

The freedom to determine family size, like the other great human freedoms, is one that history shows individuals use responsibly. Individual decisions concerning the means of fertility regulation are often as good, and commonly better, than society's institutionalized expertise. In the current world of change, rapid adjustment is necessary. The necessary adaptation in fertility regulation is beginning, but unfortunately the process has been retarded by community attitudes that tend to uphold a *status quo* in sexual behaviour, block access to contraception and sterilization, and misunderstand the role of abortion. At present, a consensus about the ethics of fertility limitation would prove more important than any improvement in the clinical aspects of family planning technology, and fresh leadership from a Pope or an Ayatollah would prevent more births than any new IUD or implantable steroid. Important decades have been wasted. Extension of family planning choices to all people as rapidly as possible may be the most important freedom to battle for in the present decade.

Family planning is like a vine. Beside the other disciplines of curative and preventive medicine, it will never be a sturdy or impressive plant, but given the right framework of social and medical policies and given the right conditions, it can produce a necessary and wanted harvest covering great areas and changing and improving health and economic circumstances in many tangible ways. At the same time, it can easily be destroyed by ring-barking with restrictive laws, unrealistic medical practices or moralizing. Success or failure in the next 10 years can help determine if our children will live in a world with 8000 million or 12000 million other people.

Suggested further reading

The economic rationality of high fertility: an investigation illustrated from Nigeria Survey Data. J. Caldwell. *Population Studies*, **31**, 5–28 (1977).

Thailand's continuing reproductive revolution. J. Knodel, N. Debvavalya and P. Kamnuansilpa. *International Family Planning Perspectives*, **6**, 84–97 (1980).

The unmet need for birth control in five Asian countries. C. F. Westoff. *International Family Planning Perspectives*, **4**, 9–18 (1978).

History of contraception. M. Potts. In *Gynecology and Obstetrics, vol. VI, Fertility Regulation*. Ed. J. J. Sciarra, M. J. Daly and G. I. Zatuchni. Lippincott; Philadelphia (1982).

The Global 2000 Report to the President of the US. G. O. Barney. Pergamon Press; New York (1980).

Rural Health and Birth Planning in China. Pi-chao Chen. International Fertility Research Program; North Carolina (1981).

Society and Fertility. M. Potts and P. Selman. Macdonald & Evans; Plymouth (1979).

Textbook of Contraceptive Practice (2nd ed.). M. Potts and P. Diggory. Cambridge University Press (1983).

Induced Abortion: A World Review, 1981. C. Tietze. Population Council; New York (1981).

Population and History. E. A. Wrigley. Weidenfield & Nicolson; London (1969).

Abortion: Medical Progress and Social Implications. Ciba Foundation Symposium 115. Pitman Press; London (1985).

4

Benefits and risks of contraception

M. P. VESSEY

Until the late 1950s, couples wishing to prevent conception had either to avoid sexual intercourse or to rely on one of a number of simple birth control methods – use of the 'safe period' or withdrawal, or of a condom or diaphragm. There was little scientific interest in these contraceptive techniques and the provision of instruction in their use was regarded as beneath the dignity of many members of the medical profession. In the last two decades, however, with the wide availability of oral contraceptives ('the Pill') and intra-uterine devices (IUDs), together with a big upsurge in interest in abortion and sterilization, research and practice in the human fertility control field have come to occupy a prominent place in medicine.

In assessing the value of any method of contraception, the answers to three major questions must be sought. How effective is the method? How acceptable is the method? How safe is the method? Before the introduction of the Pill and IUDs, the small amount of contraceptive research conducted by doctors and sociologists was concerned only with the first two of these questions. The newer methods of fertility control, however, all have effects over and above their contraceptive one and hence the third question has assumed great significance. It is important to remember that contraceptives are used by healthy young people over prolonged periods of time, and that very high standards of safety are therefore required – much higher, for example, than when dealing with a drug that might be of value in treating an older person with a crippling or possibly fatal disease.

Before considering the benefits and risks of individual contraceptive methods, I plan to discuss some of the approaches that are used in assessing effectiveness, acceptability and safety.

Assessment of contraceptive methods
Effectiveness
In general, the effectiveness of a contraceptive method is more easily studied than its acceptability or safety. Careful follow-up of a representative sample of, say, 1000 sexually active women of proven fertility using the contraceptive over a period of a year will usually give a reasonably good estimate of the accidental pregnancy rate, although if at all possible such a study should be replicated in a number of centres. Longer-term follow-up may sometimes be necessary since some contraceptive methods (e.g. 'medicated' IUDs) lose effectiveness with prolonged use. Special care has

121

to be taken in assessing any pregnancies coming to light early on in the follow-up period to try to exclude the possibility that the women concerned were already pregnant before starting to use the contraceptive. It is also important to ensure that the participants do not use any method of contraception in addition to the one under study. In one British investigation, for example, close questioning revealed that about 10 per cent of couples taking part in a trial of an IUD were using other birth control methods as well.

A distinction is often made between 'theoretical effectiveness' and 'use-effectiveness' of a contraceptive. The first of these terms relates to the efficacy of a method if it is always used exactly in accordance with instructions (whether or not this is so may, of course, be extremely difficult to determine!). The second relates to the efficacy of a method as it is actually used in practice. Clearly, as far as IUDs and injectable contraceptives are concerned, theoretical effectiveness and use-effectiveness are closely similar, since there is little opportunity for these methods to be abused. The same, however, does not apply to any method requiring continued motivation on the part of the user. Thus, for example, the theoretical effectiveness of combined oral contraceptives approaches extremely close to 100 per cent. In practice, however, somewhere between 2 and 20 women in every 1000 using the preparations experience an unwanted pregnancy in any given year, mainly because some do not take the tablets as directed. As might be expected, the use-effectiveness of methods of fertility control such as the Pill or the diaphragm varies considerably from study to study, depending on the characteristics of the participants. Nonetheless, the concept of use-effectiveness has, in general, proved of more value in assessing the efficacy of methods of fertility control than has that of theoretical effectiveness.

Use-effectiveness is commonly expressed in terms of an accidental pregnancy rate per 100 woman-years of exposure, the so-called Pearl Index (named after Raymond Pearl who described this analytical approach in 1932). In this method of computation, the exposure of, say, 100 women for 6 months each is regarded as equivalent to the exposure of, say, 50 women for 12 months each. This is unsatisfactory because, in general, higher failure rates tend to occur in the early months of contraceptive use than in the later months since the highly fecund and the highly feckless are soon discovered! Furthermore, computation of the Pearl Index can, on occasion, give rather ridiculous results. For example, in a disastrous study of the post-coital use of progestogens with which I was involved, no fewer than 26 out of 52 women became pregnant within 3 months of starting to use the method, yielding a Pearl Index of over 200!

Clearly, more sophisticated methods of analysis are desirable and the 'life table' approach, which yields cumulative rates (e.g. of accidental pregnancy or of continuation of use) per 100 women by the end of a given period of time seems to be the best available. The value of the life table approach (which had previously been applied only in mortality studies) was recognized by Robert Potter in 1963 and brought into prominence by Christopher Tietze in 1967 for the analysis of data collected during trials of IUDs. Table 4.1 illustrates the type of information made available by life table analysis; the reader who is interested in the technical aspects of the computations will find an appropriate reference at the end of the chapter.

Acceptability

Assessment of the acceptability of a contraceptive method is difficult. One objective measure is provided by finding out what proportion of a sample of subjects will agree to give a particular method a try in competition with other available methods (the *initial* acceptability of the method). A second is provided by calculation (by life table methods) of the proportion of subjects adopting a particular method who continue to use it at certain intervals of time (say, 1 year, 2 years, and so on) after starting (see Table 4.1 for example).

   Some of the obvious factors influencing the acceptability of a method of contraception are effectiveness, safety (including the occurrence of minor side-effects as well as major hazards), cost, simplicity in use and aesthetic appeal. In addition, however, characteristics of the subjects using

Table 4.1. *United Kingdom Copper 7 IUD study; cumulative event rates by type of event at selected months of use per 100 users*

Event	Ordinal months of use				
	3	6	12	18	24
Pregnancy	0.5	0.9	2.5	3.0	3.2
Expulsion	7.8	9.4	11.0	11.9	12.0
Removal					
Bleeding or pain	2.0	3.5	6.4	8.8	11.2
Other medical reasons	0.3	0.5	1.0	1.5	1.5
Planning baby	0.3	1.3	4.2	7.4	11.4
Other personal reasons	0.5	1.1	1.4	2.2	2.8
Investigators choice	0.2	0.5	0.7	1.0	6.7
Released from follow-up	1.4	2.3	4.0	5.1	6.0
Lost to follow-up	0.8	0.9	8.5	22.4	29.8
Estimated proportion continuing with device	95.7	91.7	84.1	77.1	70.9
Number of women remaining	2359	1968	1265	505	162
Woman-months of observation	7645	13914	23322	28154	29777

(Adapted from M. Williams and R. Snowden. *Family Planning Association Medical Newsletter*, No. 55, pp. 1–4 (January 1975).)

the method such as age, social class and stage of family building are extremely important. Where a method requires medical supervision (as with the oral contraceptives) the enthusiasm of the doctor or nurse concerned may also have a profound influence. From these considerations, it is clear that the acceptability of a method of contraception will vary widely from one population group to another.

Recently, Tietze has combined the concept of continuation of use with that of use-effectiveness by introducing a measure known as 'extended use-effectiveness'. This has made possible measurement not only of failures directly attributable to the use of a specific method, but also of failures that follow any discontinuation of the initial method during the period of observation, whether or not another method is substituted for it. Extended use-effectiveness is thus concerned with the important question of how many unplanned pregnancies are likely to occur during a given time period if a certain number of couples decide to adopt a particular method of contraception. Some data from Sri Lanka shown in Table 4.2 illustrate the superiority of IUDs over oral contraceptives in that country in regard to extended use-effectiveness.

Safety

In discussing approaches to the study of the safety of different methods of contraception, for the sake of simplicity, I plan to refer mainly to oral contraceptives. However, many of the techniques that I shall describe can be (and have been) applied to the investigation of other birth control methods (and, of course, to many different drugs used in medicine).

Side-effects of oral contraceptives may, for practical purposes, be divided into two broad categories – those that are common but minor, and those that are rare and serious. Two important questions arise in connection with common side-effects of the Pill (such as headache, breast tenderness

Table 4.2. *Sri Lanka contraceptive study. In this study, use-effectiveness of oral contraceptives and IUDs was about the same, i.e. about two pregnancies per 100 woman-years of use*

	Initial method of contraception	
	Oral	IUD
Continuation rate of initial method at 24 months (%)	30	68
Continuation rate of any method at 24 months (%)	47	75
Pregnancy rate at 24 months (%)	44	17

(Adapted from N. H. Wright, personal communication (1975).)

and nausea). The first of these is concerned with the *nature* of the symptoms: to what extent are they pharmacological and to what extent psychological? The only way to answer this question conclusively is the classical way, i.e. to set up a randomized 'double-blind' clinical trial in which one group of subjects is assigned to the oral contraceptive tablet and the other to an identical dummy tablet (neither the subjects nor the staff knowing which is which). Obviously, such studies present considerable problems, but it is nonetheless possible for them to be conducted on volunteers who are unexposed to the risk of pregnancy, who are willing to accept the risk of pregnancy or who are reliably protected by some other method of birth control (e.g. the condom) which is free from side-effects. Not surprisingly, few such studies have, in fact, been completed. One of the best known, that by Joseph Goldzieher in Texas, strongly suggests that only a modest proportion of the minor side-effects often attributed to oral contraceptives have a pharmacological basis.

The other major question relating to common side-effects concerns the *relative incidence* of symptoms associated with different dosages, formulations and so on. To answer this question satisfactorily also requires randomized double-blind trials, but now there is no need to have a dummy treated group.

While the methodological problems involved in evaluating common side-effects are fairly straightforward, the question of obtaining adequate information about rare major hazards (such as heart attack or cancer) is much more difficult. Table 4.3 lists the methods of investigation that are available, arranged in the order in which they can first be applied during the development of a new product.

First on this list comes *animal studies*. Such studies are, of course, of vital importance in the development and initial testing of all drugs, but it is, nonetheless, true that many rare adverse effects of drugs are not predictable on the basis of animal toxicology. Typical of the confusion that may occur is the controversy over the proper interpretation of the results of experiments that have shown that some progestogens cause tumours (of

Table 4.3. *Methods of obtaining information about rare major hazards of oral contraceptives*

1 Animal studies
2 Human studies
   (*a*) Laboratory studies
   (*b*) Morbidity and mortality studies
      (i)   Information obtained during clinical trials
      (ii)   Prospective epidemiological studies
      (iii) Case reports
      (iv)   Case-control epidemiological studies
      (v)   Vital statistics

which a proportion are malignant) to appear in the breasts of beagle bitches. These findings led to the disappearance from the market of a number of oral contraceptive preparations, and yet the British Committee on Safety of Medicines has recently come to the conclusion that the beagle is an inappropriate species for the study of the possible carcinogenic effects of contraceptive steroids! Again, as will be discussed later in this chapter, a great deal of anxiety about the safety of oral contraceptives relates to the occasional occurrence of heart attacks and other cardiovascular catastrophes in women using the Pill. To the best of my knowledge, no animal work conducted to date has been of any value in increasing our understanding of these problems.

Second on the list comes *laboratory studies in human beings*. Experience with the existing oral contraceptives has shown that relatively few subjects, say 100, are sufficient to provide reasonably comprehensive information about the effects of the preparations on liver function, blood clotting mechanisms, blood fats and so on. Studies of this type have clearly shown that oral contraceptives have an enormous variety of 'metabolic' effects. In general, however, the importance of the effects observed, in terms of whether or not they will eventually result in overt disease, is unknown.

Animal studies and laboratory studies in human beings, then, while pointing to areas where danger might lie, do not measure what we are really interested in – that is, the occurrence of illness and death attributable to the agent under study. The remaining methods of investigation shown in Table 4.3 tackle this problem directly. First of all, there is an opportunity to study major morbidity and mortality during ordinary *clinical trials*, the main purpose of these trials being, of course, to study effectiveness, acceptability and minor side-effects. Usually, however, these trials are too small and the duration of follow-up is too short for them to contribute much information about safety. Once a contraceptive agent has passed the clinical trial stage and has been released for use in the general population, *prospective epidemiological studies* may be undertaken. Studies such as these involve the long-term follow-up of large numbers of subjects using the contraceptive under investigation, together with suitable control groups. Although such prospective studies of oral contraceptives could have been started as early as 1962, it was not, in fact, until 1968 that three major investigations were launched, two in the United Kingdom and one in the United States. The first of these investigations, the Royal College of General Practitioners (RCGP) Oral Contraceptive Study, initially involved some 46000 women, recruited and followed up by about 1400 general practitioners. An interim report was published in 1974 and many of the women are still under observation. The second of the studies, the Oxford Family Planning Association (Oxford–FPA) Contraceptive Study, enlisted the aid of 17000 women attending 17 large family planning clinics. A variety of follow-up methods have been used in this study, data being collected at the clinics, by post, by telephone and at home visits. An interim report was published in 1976 and this study too is still continuing. The third study, involved about 16 500 women who were members of the Kaiser Foundation Health Plan in Northern California and who joined the study by having a general health check-up at Walnut Creek. Again, follow-up information has been obtained in a variety of ways – at return health check examinations, by post and by telephone – and a detailed report was published in January 1981. These prospective studies have been much more successful than was anticipated and have provided comprehensive information about the benefits and risks of Pill use.

With the widespread adoption of a contraceptive agent, the three remaining methods of investigation listed in Table 4.3 become of practical value. First, doctors prescribing the agent should be encouraged to *report suspected adverse reactions*, either by writing letters to medical journals or by notifying the official agencies that have been set up in a number of countries to collect such information (the Committee on Safety of Medicines in the United Kingdom). In general, case reports do not provide definite evidence of a cause-and-effect relationship, but they do serve as pointers towards areas where closer study might be useful. The initial alarm that oral contraceptives might occasionally cause certain cardiovascular catastrophes was, for example, sounded by case reports made by astute clinicians. In some circumstances, it may be possible to investigate the relationship between a drug and a suspected adverse effect by alternately administering and withdrawing the drug and seeing whether there is any corresponding variation in the intensity of the reaction. Case reports of this sort are clearly of special significance; indeed, some of the most compelling evidence that oral contraceptives can cause high blood pressure has been obtained from investigations of this type (Fig. 4.1).

Fig. 4.1. Effect of two courses of oral contraceptive on blood pressure. Prior to 1963 the woman's blood pressure had been noted to be normal. A normal pregnancy and delivery occurred in 1965–66. (Treatment with chlorothiazide and reserpine was intended to prevent blood pressure rising too high.) (From M. H. Weinberger *et al. Ann. Int. Med.* **71**, 891–902 (1969).)

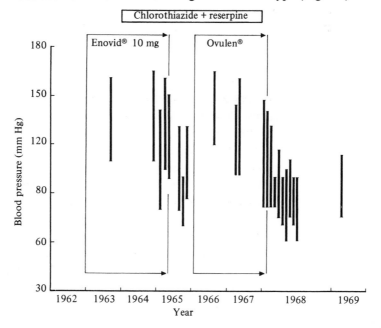

Secondly, *retrospective (case-control) epidemiological studies* of particular suspected adverse reactions can be undertaken. Since no one can hope to understand the literature on the safety of contraceptive methods without knowing what a case-control study is, it seems worth while pausing to spell out one or two of the basic principles here. In a case-control study, the starting point is the identification of an unselected group of subjects newly presenting with the disease of interest (say, with deep venous thrombosis – a condition in which blood clots form in the deep veins, usually of the leg). These subjects are then interviewed to find out how many of them were exposed to the suspected causal factor (in this example, oral contraceptives). The frequency of exposure in these 'cases' is then compared with that in an appropriate series of control subjects free from the disease of interest. The control series is, of course, designed to provide an estimate of the exposure rate that would be expected in the case series if there were no association between the exposure and the illness of interest. As an example of the sort of data that may be derived from case-control studies, Table 4.4 shows some results published by Richard Doll and me in 1968, strongly suggesting a positive association between oral contraceptive use and deep

Table 4.4. *Affected and control patients classified by use of oral contraceptives during month before onset of disease*

Diagnostic group	Number of patients	
	Oral contraceptives used	Oral contraceptives not used
Venous thrombosis and embolism	26	32
Control	10	106

(Adapted from M. P. Vessey and R. Doll. *Br. Med. J.* **2**, 1–4 (1968).)

vein thrombosis. That case-control studies are capable of providing reliable results has been shown by a large number of investigations, many concerning fields other than fertility control – for example, by studies of the relationship between German measles and congenital malformation, and between smoking and lung cancer. Case-control studies are, however, liable to distortion from bias in both the selection of the diseased individuals and of the controls, and they cannot provide much information about the size of any risk they demonstrate. Furthermore, they are designed to test specific hypotheses and cannot be expected to detect a completely unknown hazard. To compensate for these disadvantages, case-control studies can provide results quickly and, in comparison with prospective studies, their cost is low.

Finally, *vital statistical data* may yield useful information. Thus, many countries collect detailed mortality and cancer registration statistics, and examination of trends concerning disorders suspected as adverse reactions to a contraceptive agent may throw light on the existence or otherwise of an association. Evidence of this type is, however, indirect and must be interpreted with great caution.

Benefits and risks of contraceptive methods

In the following section, an attempt is made to review in a concise way modern thinking about the major benefits and risks of the main contraceptive methods. A brief reference is also made to male and female sterilization and to abortion. It is important to stress that the discussion is angled from the Western point of view. The contraceptive needs of the developing world are considered by Malcolm Potts in the preceding chapter.

Coitus interruptus

Coitus interruptus (withdrawal of the penis from the vagina just prior to ejaculation) is an ancient method of fertility control which is still very widely used. As Malcolm Potts has pointed out, it is known by a number of euphemisms and is sometimes not explicitly recognized by the user as a method of birth control. In answer to the question 'Do you use any form of contraception?', a woman will often say 'No', but if the question is framed 'Is your husband careful?', the answer will be 'Oh, yes'.

Coitus interruptus requires neither prior preparation nor medical supervision and costs nothing. Not surprisingly, there are few reliable data on efficacy, but use–effectiveness pregnancy rates are probably in the range of 10–30 per 100 woman-years. Despite popular opinion (among doctors as well as amongst lay people), there is little evidence that the practice is harmful for those who find it acceptable.

The rhythm method

The rhythm method, which is based on the fact that only a limited part of the menstrual cycle is associated with the possibility of conception, is the only fertility control method officially approved by the Roman Catholic Church. In the opinion of some, this is its sole advantage! In general, the efficacy of the method is probably similar to that of coitus interruptus, but it may be improved by the application of one or more of the techniques that give some indication of the time of ovulation, i.e. recording menstrual cycle-lengths, taking daily basal body temperatures and, more recently, daily monitoring of the 'tackiness' of the cervical mucus (Fig. 4.2). There are no proven harmful effects of the rhythm method in those who find it acceptable. Accidental pregnancies following failure are, however, particularly likely to occur very early or very late in the fertile period of the menstrual cycle, and it has been suggested that this may predispose to extra-uterine pregnancy and to fetal malformation.

The condom

The condom is by far the most widely used birth control device. Use-effectiveness pregnancy rates vary widely in different studies, but are generally within the range of 2–20 per 100 woman-years.

Among those who find the condom acceptable – and there are many who do not – there are no important harmful effects. Men who are allergic to rubber may, however, develop a rash on the penis or scrotum. Women may show allergic symptoms such as vaginitis. In such cases, the condom should be abandoned and another contraceptive method substituted.

An important medical advantage of the condom is that its use lessens the risk of transmission of venereal disease, especially gonorrhoea.

The diaphragm

Diaphragms have never been as widely used as condoms and, until recently, their popularity was on the decline. Anxiety about the safety of oral contraceptives (and, to some extent, IUDs as well) has, however, led to a resurgence of interest in the diaphragm, especially among the young, in a number of developed countries. As with the condom, reported use-effectiveness pregnancy rates show wide variation, the range being of the order of 2–25 per 100 woman-years. There is no doubt, however, that highly effective contraception can be achieved by motivated couples making regular use of the diaphragm (Fig. 4.3).

Rubber allergy may occur in women using diaphragms, or in their consorts. There is also some evidence that the use of a diaphragm may

Fig. 4.2. Chart showing rise of body temperature on day 15 (arrow) indicating that ovulation probably occurred on day 13 or day 14. (From R. L. Kleinman, ed. *Family Planning Handbook for Doctors*. IPPF (1980).)

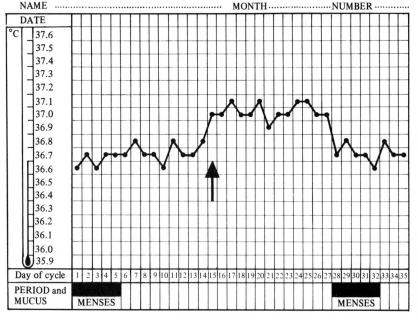

predispose to urinary tract infection in some women, presumably as a result of the pressure of the rim of the device on the urethra. On the benefit side, a number of studies have suggested that regular use of a diaphragm offers some protection against cancer of the cervix (a disease that is closely related to sexual activity, occurring predominantly in women who have had intercourse with multiple partners, particularly at an early age). Perhaps not surprisingly, the risk of infections of the uterus, Fallopian tubes and surrounding structures (often loosely termed 'pelvic inflammatory disease') seems to be reduced in regular diaphragm users as well. This is an important beneficial effect because pelvic inflammatory disease can be followed by permanent infertility due to occlusion of the Fallopian tubes by adhesions (see Chapter 5).

Vaginal spermicides

Vaginal spermicides are available as pessaries, jellies, creams and foams. They destroy spermatozoa by a detergent action and should always be used in conjunction with an occlusive method of birth control. When used alone, use-effectiveness pregnancy rates in the range of 10–40 per 100 woman-years are to be anticipated.

Recently, there has been some anxiety about the safety of spermicides. It has been shown that these substances are absorbed into the blood stream through the vaginal wall, and questions have been raised about the possible occurrence of systemic toxic effects, such as damage to the liver. No such toxic effects have, however, been discovered to date. There is also concern about possible harmful effects of spermicide use on the outcome of pregnancy, particularly in the event of contraceptive failure. Fertilization might then involve 'blighted' spermatozoa, or the developing embryo might be damaged by absorbed spermicide. Some evidence has been presented that spermicide users do, in fact, have a slightly increased risk

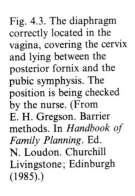

Fig. 4.3. The diaphragm correctly located in the vagina, covering the cervix and lying between the posterior fornix and the pubic symphysis. The position is being checked by the nurse. (From E. H. Gregson. Barrier methods. In *Handbook of Family Planning*. Ed. N. Loudon. Churchill Livingstone; Edinburgh (1985).)

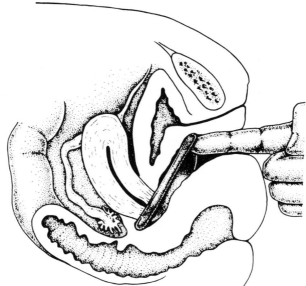

of experiencing spontaneous abortion or giving birth to a malformed child, but the data are far from conclusive.

Steroidal contraceptives

The basic physiological principles underlying a hormonal approach to contraception had already been elaborated by the mid-1930s, but the development of practical hormonal methods of birth control had to await the synthesis of potent orally active steroids some 20 years later. Much of the physiological and clinical development work in producing 'the Pill' was done by M. C. Chang, Gregory Pincus and John Rock in the United States in the 1950s and great credit must be given to these three for their contribution to one of the great 'medical breakthroughs' of the twentieth century. Indeed, in 1981 it was estimated that about 55 million women around the world were taking oral contraceptives, although in recent years sales have levelled off, or declined to some extent, in Western Europe and the United States (where use rates are now mostly in the range of 15–30 per cent of women aged 15–44 years).

Combined oral contraceptives. Combined preparations contain an oestrogen and a progestogen, and are usually given for 21 days, followed by a 7-day break, during which uterine 'withdrawal' bleeding occurs. They act primarily by causing a diminution in the output of pituitary gonadotrophins, leading to an inhibition of ovulation. In addition, however, they may interrupt fertility by effects on the ovary, tubes, endometrium and cervical mucus. At present, the use of combined oral contraceptives represents the most effective reversible method of fertility control that is widely available. Thus, use-effectiveness pregnancy rates in different studies have generally been within the range of 0.2–2.0 per 100 woman-years.

Until the mid-1970s, most of the available data about the benefits and risks of the Pill had been derived from uncontrolled clinical trials and from case-control studies. Since then, an enormous amount of information has been obtained from the large prospective studies referred to on pages 126–7. Data about the effects of the Pill in developing countries are, however, extremely sparse, although a number of agencies, such as the World Health Organization, are trying to conduct appropriate studies. In the meantime, the reader is cautioned *not* to extrapolate the data summarized here to parts of the world to which they clearly do not apply!

Turning now to the *beneficial effects* of the combined Pill, pride of place must be given to its remarkable efficacy, which, coupled with a high degree of acceptability (at least among the young in the developed world), has given many women a new freedom from anxiety about unwanted pregnancy.

Apart from this, however, oral contraceptives have a number of other well-documented important beneficial effects. First, they tend to suppress some menstrual disorders, notably excessive bleeding (menorrhagia) and period pain (dysmenorrhoea), leading to a reduction in hospital admissions for investigation and treatment of these problems, and to a lessened risk of iron-deficiency anaemia. Secondly, case-control and prospective studies have consistently shown that long-term Pill users develop fewer benign breast lumps than other women. This is an important benefit because lumps in the breast often have to be surgically biopsied to exclude the possibility of cancer being present. Thirdly, since oral contraceptives act principally by inhibiting ovulation, it is not surprising that functional ovarian cysts (i.e. follicular cysts and corpus luteum cysts) are relatively uncommon in Pill users. While such cysts usually pass undetected and are of no consequence, from time to time they give rise to an abdominal mass demanding investigation, or they may rupture causing acute abdominal pain and requiring surgical intervention. Finally, and perhaps most important of all, epidemiological studies reported during the last 5 years have shown that the risk of both ovarian cancer and cancer of the endometrium is reduced by about 50 per cent in women who have used the combined Pill for at least 2 or 3 years. Furthermore, this protective effect appears to persist for many years after cessation of Pill use.

While the beneficial effects already described may be considered established, a number of others that have been reported in some studies also deserve mention. These include a lessened risk of thyroid disease, rheumatoid arthritis, peptic ulcer disease and pelvic inflammatory disease. Further work is necessary before the significance of these observations can be assessed, but their potential importance is obvious.

Before discussing the *major risks* of the combined Pill, it should be noted that oral contraceptives are known to cause minor side-effects such as nausea, headache and breast tenderness (see pages 124–5). Although such symptoms are common enough and troublesome enough to lead to discontinuation of the Pill by up to 25 per cent of women, they disappear as soon as treatment is stopped and so do not, in my opinion, represent a 'risk'. Rather, in the following paragraphs, I am concerned with adverse effects of the Pill of sufficient severity to require hospital investigation or treatment (and which may, occasionally, prove fatal) and with effects on fertility and on the unborn child.

The best-known adverse effects of oral contraceptive use are the cardiovascular ones. These effects were first suggested by case reports submitted by clinicians, were then investigated by means of case-control studies (see page 128) and were finally confirmed when the results of prospective studies became available. It now seems clear that the Pill can cause deep vein thrombosis (and its most serious complication, pulmonary embolism, which occurs if a piece of the thrombus becomes detached and is carried through the heart to be impacted in the pulmonary arteries), certain types of stroke, and heart attack (acute myocardial infarction – death of heart tissue from blockage of blood vessels). There is reasonably convincing evidence that all these hazards are positively related to the dose of oestrogen in the Pill and this is the main reason why the oestrogen content of oral contraceptives has been progressively reduced over the years by the pharmaceutical companies. The progestogenic component may be of importance too, especially in relation to stroke and heart attack. None of the vascular hazards appears to be strongly related to duration of Pill use, and the increased risks of venous thrombosis and heart attack seem to disappear after the Pill is stopped. The risk of stroke may, however, persist in ex-users of the Pill. Results obtained during the last few years have shown that fatal adverse cardiovascular reactions to the Pill are very rare in women under 35 years of age and are strongly concentrated in those who smoke. This point is taken up again later in the chapter during the discussion of the balance of benefits and risks. The mechanisms underlying adverse cardiovascular reactions to the Pill are uncertain, but it is known that the preparations have considerable effects on blood pressure levels, blood fats and the blood coagulation system.

Women taking oral contraceptives are at a slightly increased risk of suffering from gallstones requiring surgical treatment. The preparations also have an effect on the risk of certain benign liver tumours, especially so-called hepatocellular adenoma. This condition, which may have a fatal outcome, is extremely rare in women of childbearing age: in those without exposure to the Pill, the incidence is around one per million per annum. Case-control studies in the United States have shown that oral contraceptive users suffer a much higher incidence than this, the risk being very strongly correlated with the steroidal content and duration of use of the preparations on the one hand, and the age of the user on the other. Some workers suspect that oral contraceptives may also increase the risk of liver cancer, but the evidence is not conclusive.

Cervical erosion (in which a zone of columnar epithelium develops on the vaginal portion of the cervix uteri in place of the stratified squamous epithelium normally found there) occurs about twice as often in Pill users as in others. Although a trivial condition in itself, a diagnosis of cervical erosion may lead to hospital admission for cautery treatment under general anaesthetic.

After discontinuing the Pill, some women take a few weeks to restart normal menstruation, but there is no convincing evidence that prior pill use is a cause of prolonged secondary amenorrhoea (absence of periods for more than 6 months). Prospective studies have shown, however, that many women do experience some temporary impairment of fertility after stopping the Pill (Fig. 4.4). In the majority, this lasts only a couple of months, but in some, recovery may be much slower. It seems unlikely that oral contraceptives are ever a cause of permanent infertility.

Many epidemiological studies have been concerned with the possible relationship between oral contraceptive use and malignant disease. With regard to ovarian and endometrial cancer, I have already mentioned that there is clear evidence of a protective effect of combined preparations.

Large case-control studies conducted during the 1970s have shown that there is no *general* association between oral contraceptive use and breast cancer, but work recently reported from Los Angeles has suggested that prolonged early use of the Pill (before the age of 25 years) may increase the risk of the disease. This finding has not, however, been confirmed by the vast Cancer and Steroid Hormones (CASH) Study conducted by the Centers for Disease Control in the United States, so the issue must be regarded as unresolved. Further work is urgently needed (and has, in fact, been started in the United Kingdom) because breast cancer is a common disease which carries a high mortality. There is also some recently published evidence that prolonged oral contraceptive use may increase the risk of pre-invasive and invasive cancer of the cervix, but once again, the association cannot be regarded as proved. Indeed, since cancer of the cervix is strongly associated with sexual activity, it is extremely difficult to detect any independent effect of the method of contraception used.

Ex-users of oral contraceptives who become pregnant are not at any increased risk of an unfavourable outcome of pregnancy. A number of reports have, however, suggested that oral contraceptives taken inadvertently during pregnancy may slightly increase the risk of malformation of the fetus (especially heart defects, neural tube defects and limb development defects). The evidence is, however, sparse, uneven and difficult to interpret. In my opinion, the risk certainly cannot be regarded as substantiated.

Finally, many other disorders, including migraine, depression, and

Fig. 4.4. Fertility after stopping different methods of contraception in order to conceive. (From M. P. Vessey *et al. Br. Med. J.* **i**, 265–7 (1978).)

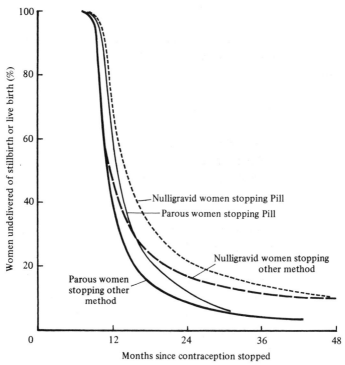

urinary tract infection have been reported in some studies, but not in others, as possible adverse effects of oral contraception.

Progestogen-only oral contraceptives. Low doses of progestogens taken every day by mouth have been extensively investigated as contraceptives. Such preparations do not consistently inhibit ovulation and their mode of action is uncertain. Their efficacy is lower than that of the combined Pill; use-effectiveness pregnancy rates in different studies have generally been in the range 2–10 per 100 woman-years. Another major disadvantage of progestogen-only oral contraceptives is their tendency to disrupt the menstrual cycle in many women, producing irregular bleeding. These drawbacks probably account for the fact that progestogen-only Pills represent well below 10 per cent of all oral contraceptives consumed. The main advantage of progestogen-only Pills is that they appear to be largely free from the undesirable metabolic effects (e.g. on carbohydrate and fat metabolism) of combined preparations.

Injectable steroidal contraceptives. The most widely studied injectable steroidal contraceptive is medroxyprogesterone acetate (Depo-Provera) given in a dose of 150 mg every 3 months. This regimen is highly effective, comparing favourably with the use of combined oral contraceptives. The principal mode of action appears to be suppression of ovulation.

Unfortunately, this method of fertility control has a number of major disadvantages. First, the normal menstrual cycle is usually disrupted, giving rise initially to irregular bleeding and later, in many women, to amenorrhoea. Secondly, menstruation, ovulation and fertility are often slow to return after the injections are stopped, although it appears that these functions return within a year in most subjects. Thirdly, medroxy-progesterone acetate has produced somewhat disturbing findings in some animal toxicological experiments, i.e. tumours in the breasts of beagle bitches, and, more recently, two cases of endometrial cancer among 16 rhesus monkeys given 50 times the equivalent human dose for 10 years! At the time of writing, Depo-Provera was not licensed for general use as an injectable contraceptive in the United States, but the British Minister of Health had announced that a product licence was to be granted in the United Kingdom to enable the drug to be used 'where no other contraceptive method is suitable'.

Despite its disadvantages, Depo-Provera has achieved considerable popularity in some parts of the world. Indeed, there is evidence that the drug increases milk volume in lactating women without affecting milk quality, an important issue in countries where the survival of children is closely associated with the success of breast feeding. There remains, however, some concern over the possible consequences of transfer of the steroid to the infant in the milk.

Post-coital contraception. Recently, there has been an upsurge in interest in the use of oestrogen–progestogen formulations as 'post-coital' contraceptives. In the United Kingdom, 100 μg ethinyloestradiol and 500 μg laevonorgestrel is usually given within 72 h of intercourse and repeated 12 h later. This approach should be considered only in emergencies and is not recommended for routine birth control. The way in which post-coital contraception works is uncertain, but it appears to be highly effective.

Intra-uterine devices

The observation that a foreign body placed in the uterus causes infertility is an old one. However, problems with pelvic infection (a serious matter in pre-antibiotic days) prevented widespread adoption of this approach to fertility control until the late 1950s. Currently, it is estimated that about 65 million women are using IUDs, of whom 50 million live in the People's Republic of China!

IUDs are manufactured in many different shapes and sizes; some of the best known are illustrated in Fig. 4.5. The Lippes loop and the Saf-T-coil are made of inert plastic and are now outmoded. The other devices represent a different approach to the problem. The principle is to use a small device, which is well tolerated by the uterus, as a carrier of a substance that alters the intra-uterine milieu by interfering with metabolic

Fig. 4.5. Some examples of intra-uterine devices. (See also Figs. 2.16 and 2.20.) (Modified from R. L. Kleinman, ed. *Family Planning Handbook for Doctors.* IPPF (1980).)

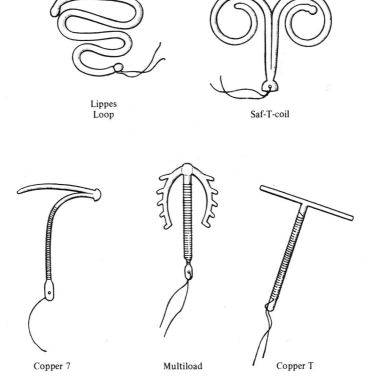

Lippes Loop   Saf-T-coil

Copper 7   Multiload   Copper T

Benefits and risks of contraception

processes in the endometrium. Copper acts well in this respect, and for this reason some of the devices illustrated carry a coil of copper wire around the stem. Progestogens, which also have a specific local anti-fertility effect, can be 'delivered' by means of IUDs as well (as in the Progestasert device). The mode of action of IUDs is still uncertain. With the inert devices,

Fig. 4.6. Frequency of pre- and post-insertion blood losses in parous women. (*a*) Lippes Loop D (*n* = 76). Pre-insertion: median 35 ml, mean 41 ml, SD 29 ml. Post-insertion: median 84 ml, mean 90 ml, SD 46 ml. (*b*) Dalkon Shield (*n* = 70). Pre-insertion: median 30 ml, mean 46 ml, SD 41 ml. Post-insertion: median 68 ml, mean 81 ml, SD 45 ml. (*c*) Copper 7 (*n* = 91). Pre-insertion: median 30 ml, mean 45 ml, SD 40 ml. Post-insertion: median 49 ml, mean 63 ml, SD 41 ml. (From J. Guillebaud *et al. Lancet*, i, 387–90 (1976).)

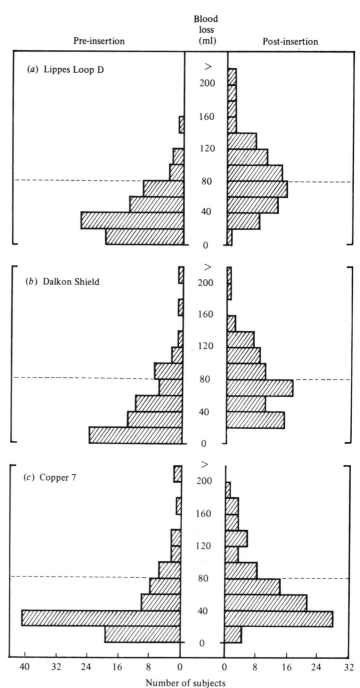

the principal mechanism appears to be the stimulation of leucocyte mobilization within the uterus which prevents the nidation of the normal blastocyst arriving from the Fallopian tube. The mode of action of copper-bearing (or hormone-bearing) devices is probably more complex. Use-effectiveness pregnancy rates of IUDs in most studies have been within the range of 1.0–3.0 per 100 woman-years.

A troublesome problem associated with the use of IUDs is their spontaneous expulsion from the uterus. For all types of device, this is most likely to occur soon after insertion; thereafter the risk drops rapidly. Sometimes the woman is unaware that expulsion has taken place and such an event is usually followed by pregnancy!

Minor side-effects commonly occur after insertion of an IUD. Of these, the most important are heavy menstrual blood loss and uterine 'cramping'. These problems tend to diminish with the passage of time. The increase in blood loss may cause anaemia in women with an inadequate intake of iron (Fig. 4.6). Bleeding and pain are the commonest medical problems leading to removal of IUDs. These abnormalities tend to be less common and less severe with the smaller 'medicated' devices than with the larger inert ones.

A number of major adverse effects may follow insertion of an IUD. First, very occasionally, uterine perforation may occur with migration of the device into the peritoneal cavity. Serious complications, such as intestinal obstruction, may then follow, especially if the device is a copper-bearing one. Secondly, pelvic inflammatory disease, which may be severe, is a well-established complication of IUD use. Thirdly, accidental pregnancies in women using an IUD are especially likely to be in the Fallopian tube, where they represent a serious medical emergency, or to end in spontaneous abortion. Furthermore, there is evidence that spontaneous abortions occurring with an IUD *in situ* are more likely than usual to be complicated by infection which, very rarely, may have a fatal outcome. The last mentioned problem caused a great furore in the United States in the mid-1970s, which led to the disappearance of one particular type of device (the Dalkon Shield) that was thought to be the chief culprit.

Although data are few, there is no evidence to suggest that IUDs have any effect on the risk of cancer of the cervix uteri or cancer of the endometrium. Congenital malformations do not appear to be more common than usual among infants born to mothers with an IUD still in place. It seems clear that IUDs must, on occasion, be responsible for permanent sterility in view of their association with pelvic inflammatory disease. No clear effect on fertility has, however, been detected in any of the studies (such as the Oxford–FPA Study) in which substantial numbers of women having a device removed in order to become pregnant have been followed up.

Sterilization

Male sterilization. Vasectomy is a simple surgical procedure which can be carried out under local or general anaesthesia. It has gained increasing acceptance in many countries during the last decade. The operation should be regarded as an irreversible method of fertility control even though re-anastomosis is sometimes successful. A man is not sterile immediately after vasectomy, as mature spermatozoa remain in the vasa deferentia and accessory glands. Accordingly, a semen specimen must be examined some weeks after the operation, and only if spermatozoa are absent can the couple abandon other methods of birth control.

Vasectomy is an extremely safe procedure and, as yet, has not been shown to have any harmful long-term effects, although minor post-operative complications are not uncommon (Fig. 4.7). A high proportion of vasectomized men, however, develop antibodies against spermatozoa, and there is some concern about the possible occurrence of 'auto-immune' conditions as remote sequelae to vasectomy. This concern was heightened by work done by Nancy Alexander in the United States which showed that atherosclerosis (the basic arterial disease process underlying most heart

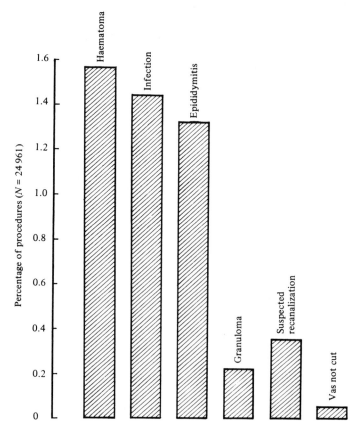

Fig. 4.7. Vasectomy complications. (Selected studies, 1969–74.) (From *Population Reports, Series D*, No. 2, Jan. 1975. Department of Medical and Public Affairs, George Washington University, USA.)

attacks and strokes) is commoner in vasectomized rhesus and cynomolgus monkeys than in control monkeys (see Book 4, Chapter 6, Second Edition). Alexander believes that this observation may have an immunological basis, but she is extremely cautious about extrapolating her findings to man. Nonetheless, a substantial number of epidemiological studies are now in progress (of both case-control and prospective design) to see whether any adverse cardiovascular effects are apparent in man. Results published to date have been entirely reassuring.

Female sterilization. Tubal ligation can be carried out via an abdominal or a vaginal incision (Figs. 4.8 and 4.9). The procedure is very safe, many large series having been reported without mortality. Such surgical

Fig. 4.8. Mini-laparotomy: a commonly used abdominal approach for sterilization. The probe in the uterus and vagina enables the oviduct to be brought close to the abdominal incision. (From I. Lubell and R. Frischer. *Proc. Roy. Soc. Lond., ser B*, **195**, 93–114 (1976).)

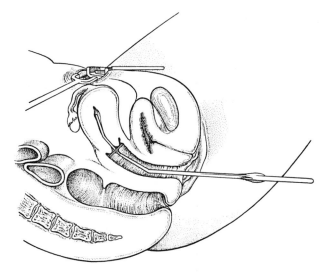

Fig. 4.9. Colpotomy: the vaginal approach for sterilization. (From I. Lubell and R. Frischer. *Proc. Roy. Soc. Lond., ser. B*, **195**, 93–114 (1976).)

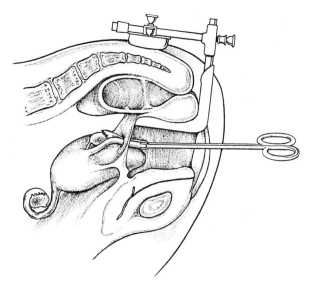

procedures, however, often require immobilization of the woman for a few days and carry with them some risk of wound infection. Accordingly, in Western countries, endoscopic methods are now most commonly used. Sterilization via a laparoscope requires hospitalization for only a day or less. During the procedure, the Fallopian tubes can be 'coagulated' with diathermy or occluded by the application of clips or bands (Fig. 4.10). The latter method seems to be displacing the former since it is safer and offers more opportunity for successful reversal should the woman subsequently regret having undergone sterilization.

All methods of tubal sterilization (like vasectomy) carry some risk of failure and perhaps 1 per cent of women will eventually experience pregnancy after the operation. The great majority of the failures reveal themselves during the first year. Pregnancies following sterilization failure are commoner than usual in the Fallopian tube.

It is often believed that some women develop menstrual problems as a result of tubal sterilization. The evidence on which this belief is based is, however, far from conclusive.

Abortion

Abortion during the first trimester of pregnancy is a straightforward procedure, the two most commonly used methods being cervical dilatation with uterine curettage, and vacuum aspiration (Fig. 4.11). There is, however, no completely satisfactory method of inducing later abortion. Hysterotomy (i.e. surgical opening of the uterus) may be used, but the operation carries an appreciable risk. Other methods include the intra-amniotic injection of hypertonic saline, urea or glucose, or the extra-amniotic or intra-amniotic instillation of prostaglandins.

Fig. 4.10. Two-puncture laparoscopy. (From I. Lubell and R. Frischer. *Proc. Roy. Soc. Lond., ser. B*, **195**, 93–114 (1976).)

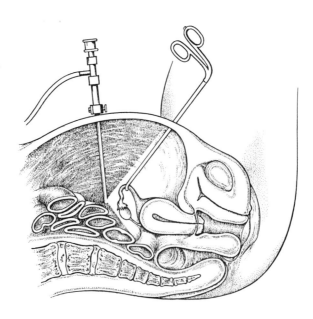

Complications of abortion may be broadly classified as short-term or long-term. Numerous studies of short-term complications (e.g. uterine perforation, cervical injury, pelvic infection, haemorrhage) have been carried out in many different countries. It is impossible to produce a concise summary of the findings because so many factors are of importance. These include the skill of the operator, the conditions under which the operation is performed, the duration of pregnancy, associated medical or surgical conditions, the technique of abortion used, the type of anaesthetic administered, whether or not sterilization is carried out at the same time, and the age, parity and socio-economic status of the woman. Taking these considerations into account, it is not surprising that the lowest death rates for induced abortion have been achieved in countries where the operation is freely available. In such countries, the mortality rates are typically between 1 and 4 per 100 000 abortions.

Possible long-term sequelae of abortion include secondary infertility and an unfavourable outcome of any subsequent pregnancy in terms of spontaneous abortion on the one hand, or premature labour and the birth of an underweight infant on the other. Despite numerous epidemiological studies of outcome of pregnancy (but *not* of infertility) in women with a past history of abortion, there seems to be no clear consensus among different investigators – probably, once again, because so many factors are of importance. However, it seems fair to conclude that the long-term

Fig. 4.11. Apparatus for uterine vacuum aspiration. (From J. Peel and D. M. Potts. *Textbook of Contraceptive Practice.* Cambridge University Press (1969).)

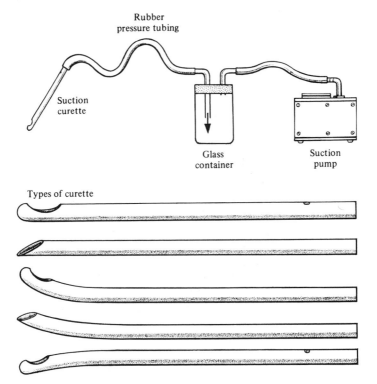

sequelae are likely to be negligible in young, healthy women when early abortion is conducted by a skilled operator using modern methods under good conditions.

Balance of benefits and risks of different contraceptive methods

From the preceding section, it will be apparent that a substantial amount of information is available about the benefits and risks of different birth control methods. Accordingly, it is now possible to draw up rough numerical comparisons of the various contraceptive techniques, and one such 'medical audit' is shown in Tables 4.5 to 4.7.

Table 4.5 is concerned with the morbidity occurring among 100000 women of childbearing age using oral contraceptives, a diaphragm or an IUD for a period of 1 year. The data are largely derived from the Oxford–FPA Contraceptive Study and concern hospital admissions. Of the conditions listed, it should be remembered that ovarian and endometrial

Table 4.5. *Morbidity associated with different reversible methods of contraception. The data are rates of hospital admission per 100000 women of childbearing age per annum.*

The data below the line relate to accidental pregnancies. It has been assumed that the failure rate for oral contraceptives is 1.5 per 1000 per year, for diaphragm users is 50 per 1000 per year and for IUD users is 20 per 1000 per year

Condition	Users of		
	Oral contraceptives	Diaphragm	IUD
Menstrual disturbance	260	300	440
Anaemia	40	60	90
Benign breast lump	120	280	280
Functional ovarian cyst	85	120	120
Ovarian cancer	3	6	6
Endometrial cancer	2	4	4
Venous thrombosis (and embolism)	90	20	20
Stroke	45	10	10
Acute myocardial infarction	10	3	3
Gallstones	180	120	120
Hepatocellular adenoma	3	0	0
Erosion of cervix uteri	540	250	250
Pelvic inflammatory disease	55	55	200
Uterine perforation	0	0	25
Term birth	90	3040	500
Spontaneous abortion	19	640	770
Extra-uterine pregnancy	1	20	120
Induced abortion	40	1300	610

cancer, the cardiovascular disorders, hepatocellular adenoma, pelvic inflammatory disease and extra-uterine pregnancy are more serious than the rest, in that they are sometimes fatal and carry a risk of permanent sequelae when they are not. The possible adverse effects of oral contraceptives on breast cancer and cancer of the cervix are not considered in Table 4.5 since, as already explained, these problems are still being evaluated.

Table 4.6 is concerned with mortality among women aged 20–34 years attributable to the use of the different methods of contraception and to the accidental pregnancies resulting from their failure. Table 4.7 corresponds exactly to Table 4.6 save that the data relate to women aged 35–44 years. The figures for cardiovascular mortality attributable to oral contraceptive use in both tables are derived from the RCGP Oral Contraceptive Study. As far as mortality is concerned, it seems reasonable to conclude that oral contraceptives compare unfavourably with the diaphragm and the IUD in women over the age of 35 years, especially in those who smoke.

Simple analyses of the sort presented in Tables 4.5–4.7, while of value in providing crude comparisons between contraceptive methods, should not be taken too seriously. Many of the estimates (especially of the risk of death) are based on very small numbers, while no allowance has been made for variations in risk with different oral contraceptive formulations, different durations of contraceptive use and so on. Furthermore, the risks associated with a contraceptive failure can be reduced to a minimum by early termination of pregnancy (carrying a pregnancy to term is more hazardous than early abortion).

Table 4.6. *Mortality (per 100 000 per annum) for women aged 20–34 years attributable to the use of different reversible methods of contraception. Contraceptive failure rates as for Table 4.5*

| | Users of | | | |
| | Oral contraceptives | | | |
Mortality	Non-smokers	Smokers	Diaphragm	IUD
Attributable to method:				
Cardiovascular disease	1.7	10.0	—	—
Ovarian and endometrial cancer	—	—	0.3*	0.3*
Pelvic inflammatory disease and uterine perforation	—	—	—	0.2
Attributable to accidental pregnancies	0	0	0.5	0.7
Total	1.7	10.0	0.8	1.2

* These figures represent the effect of failure to obtain the protection offered by the Pill.

The analyses presented apply mainly to couples concerned with the problems of birth spacing. For those who have completed their families, male sterilization is the safest method, since (as described earlier) vasectomy is associated with little morbidity and virtually no mortality. Female sterilization is preferable to the continued use of reversible methods of contraception, but the procedure is not, of course, entirely without hazard. It gives the best results in terms of morbidity and mortality if carried out at a reasonably young age, e.g. at 35 years rather than 45 years.

In closing this chapter, I would like to remind the reader of some of its many limitations! First, and most important, the discussion is entirely angled from the Western point of view – indeed, almost all the data presented have been derived from studies in the United Kingdom and the United States. Secondly, and perhaps inevitably in view of my own background, the presentation is strongly 'medical': it deals at length with the physical effects of contraceptive use and hardly touches on the psychological and social. Finally, the chapter deals only with well-established contraceptive methods in widespread use; no attempt is made to give any insight into the direction of research aimed at the discovery of new methods or the improvement of existing ones.

Table 4.7. *Mortality (per 100 000 per annum) for women aged 35–44 years attributable to the use of different reversible methods of contraception. Contraceptive failure rates as for Table 4.5*

	Users of			
	Oral contraceptives			
Mortality	Non-smokers	Smokers	Diaphragm	IUD
Attributable to method:				
Cardiovascular disease	15.1	48.2	—	—
Ovarian and endometrial cancer	—	—	2.0*	2.0*
Pelvic inflammatory disease and uterine perforation	—	—	—	0.2
Attributable to accidental pregnancies	0.1	0.1	2.3	1.4
Total	15.2	48.3	4.3	3.6

* These figures represent the effect of failure to obtain the protection offered by the Pill.

Suggested further reading

A comprehensive review of injectable contraception with special emphasis on depo-medroxyprogesterone acetate. I. S. Fraser and E. Weisberg. *Medical Journal of Australia*, Special Supplement 1, 1–19 (1981).

A placebo-controlled, double-blind crossover investigation of the side-effects attributed to oral contraceptives. J. W. Goldzieher, L. E. Moses, E. Averkin, C. Scheel and B. Z. Taber, *Fertility and Sterility*, 22, 609–23 (1971).

Breast cancer in young women and use of oral contraceptives: possible modifying effect of formulation and age at use. M. C. Pike, B. E. Henderson, M. D. Krailo, A. Duke and S. Roy. *Lancet*, ii, 929–30 (1983).

Statistical evaluation of contraceptive methods. C. Tietze and S. Lewit. *Clinical Obstetrics and Gynecology*, 17, 121–38 (1974).

Female hormones and vascular disease – an epidemiological overview. M. P. Vessey. *British Journal of Family Planning*, Supplement 6, 1–12 (1980).

Neoplasia of the cervix uteri and contraception: a possible adverse effect of the pill. M. P. Vessey, M. Lawless, K. McPherson and D. Yeates. *Lancet*, ii, 930–4 (1983).

The effects of induced abortion on subsequent reproduction. C. J. R. Hogue, W. Cates and C. Tietze. *Epidemiologic Reviews*, 4, 66–94 (1982).

Epidemiological studies of the long-term effects of birth control methods. M. P. Vessey, pp. 365–77. In *Recent Advances in Fertility Regulation*. Ed. C. C. Fen, D. Griffin and A. Woolman. WHO; Geneva (1981).

Intrauterine devices and their complications. D. A. Edelman, G. S. Berger and L. G. Keith. Martinus Nijhoff; The Hague (1979).

Risks, Benefits and Controversies in Fertility Control. Ed. J. J. Sciarra, G. I. Zatuchni and J. J. Speidel. Harper & Row; Hagerstown (1978).

Oral contraceptives in the 1980s. *Population Reports*, Series A, No. 6. The Johns Hopkins University; Baltimore (1982).

Vasectomy – safe and simple. *Population Reports*, Series D, No. 4. The Johns Hopkins University; Baltimore (1983).

Steroid contraception and the risk of neoplasia. *World Health Organization, Technical Report Series*, No. 619. WHO; Geneva (1978).

The effect of female sex hormones on fetal development and infant health. *World Health Organization, Technical Report Series*, No. 657. WHO; Geneva (1981).

5

Alleviating human infertility

*JACQUES COHEN, CAROLE FEHILLY AND
ROBERT EDWARDS*

Infertility can be defined as 'the involuntary incapacity to participate in reproduction', but for practical purposes 'infertile' usually means 'unable to conceive'. Fertility, often in the form of a goddess or a phallic symbol, has been worshipped and revered since the beginning of history (Fig. 5.1). The 'barren woman' has been a sorry figure for pity or scorn through the ages; the 'barren man' is a very recent phenomenon, consequent on our ability to diagnose infertility. Cures offered to the infertile woman in her distress have ranged from magic spells to snakebite. In an overpopulated world where some countries have introduced legislation to discourage large families, China's 'one-child' policy being the most extreme example, the plight of the childless can easily be overlooked or even dismissed as selfish. But the desire to have a child is such a fundamental and powerful human need that we cannot ignore it and still call ourselves humane.

When childbirth statistics are consulted to try to estimate the incidence of infertility, it is difficult if not impossible to distinguish true biological (involuntary) infertility from socio-economic (voluntary) infertility, the latter having a far more profound effect, especially today. There are few well-documented studies of a society sustaining maximum fertility, but the Hutterite Brethren, a religious sect living in Canada and the USA, are an excellent example (Fig. 5.2). During the period analysed, the Hutterites used no contraception, extra-marital intercourse and abortion were prohibited and the society was economically sound. Two-thirds of the women produced between seven and twelve children, the average family size being ten, and only 3 per cent of the couples were childless. The incidence of childless marriages (voluntary and involuntary) in the UK today is about 10 per cent. However, in the last 100 years the average UK family size has dwindled from six (live) children to just under two today, yet there is no real evidence to suggest a true increase in biological infertility. During the same 100-year period, general health and life expectancy have improved, so a decrease in fertility would be contrary to expectation. What we are witnessing is voluntary infertility on a huge scale.

In couples with unprotected intercourse (no contraception), 90 per cent will have achieved a pregnancy within 1 year (see Book 4, Chapter 2, Second Edition), and a few others will conceive after another year. A couple are generally regarded as infertile when unprotected intercourse has not produced a pregnancy after 2 years of trying, although 1 year's failure already gives some cause for concern, and the age of the woman should be seriously considered before delaying an investigation. If the woman has been on oral contraceptive therapy prior to the period of unprotected intercourse, the possibility of anovulatory cycles must be taken into account (see Chapter 4 by Martin Vessey).

Problems with coitus must be eliminated as a possible factor, and the frequency of sexual intercourse considered with obvious reference to the time when conception is most likely to occur. True psychogenic infertility

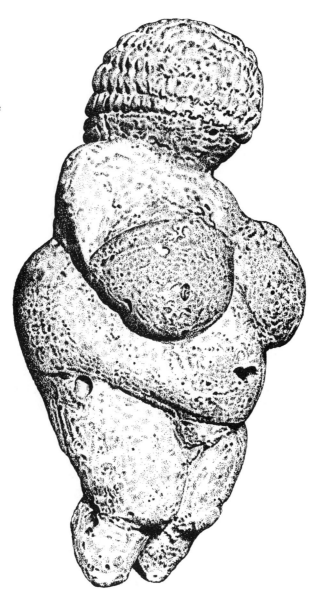

Fig. 5.1. Venus of Willendorf. Paleolithic limestone statuette, assumed to be a fertility symbol. It is one of the earliest representations of the human form, and has an undefined face but large buttocks and abdomen, and pendulous breasts. It was found in Austria in 1908. (Naturhistorisches Museum, Vienna.)

remains a possibility with little hard evidence to support it, so far. Some psychiatric conditions, notable schizophrenia, can be associated with disturbances of reproductive physiology, but this is rare. Psychological problems can interfere with reproduction, but this is due to sexual dysfunction (impotence, premature or retarded ejaculation, vaginal spasm) rather than physiological disturbance. Reports that couples undergoing infertility investigations are depressed, have low self-esteem or anxiety symptoms are hardly surprising, but it is a 'chicken and egg' situation: which came first, the infertility or the psychological symptoms?

There is a considerable variation in the incidence of different types of infertility when the data are examined from infertility clinics in different areas of the country, or between infertility clinics in different countries. Despite these differences, which may be attributed to geographical or

Fig. 5.2. The fertility of the Hutterite population is one of the highest ever recorded in any human society. The relatively low fertility of England and Wales reflects voluntary infertility rather than biological infertility. (Adapted from T. B. Hargreave. *Male Infertility*, p. 3 – see Suggested further reading.)

socio-economic factors, or simply to different diagnostic procedures, rough estimates can be made of the source of the infertility within the couple:

infertility caused by the female only 35–55 per cent
infertility caused by the male only 30–50 per cent
infertility caused by both partners 15–35 per cent.

The duration of infertility of the couple who request treatment is a good indicator of the chance of a pregnancy being established (Fig. 5.3). This is particularly true of male infertility. A few types of male infertility can be treated with moderate success, but results are poor with other types. Claims made for new treatments have rarely been well tested against a control population of infertile patients, and a pregnancy due to the normal 'background' rate of spontaneous conception is often mistaken for the results of a course of treatment.

Female infertility

Female infertility has caused great concern since ancient times. Our current knowledge of the subject is considerable: for every book on male infertility, there must be at least a dozen on female infertility, but this imbalance will perhaps be redressed when andrology becomes as well-developed a speciality as gynaecology.

The most obvious parameters for investigating female infertility involve

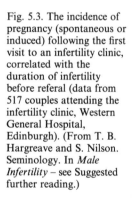

Fig. 5.3. The incidence of pregnancy (spontaneous or induced) following the first visit to an infertility clinic, correlated with the duration of infertility before referal (data from 517 couples attending the infertility clinic, Western General Hospital, Edinburgh). (From T. B. Hargreave and S. Nilson. Seminology. In *Male Infertility* – see Suggested further reading.)

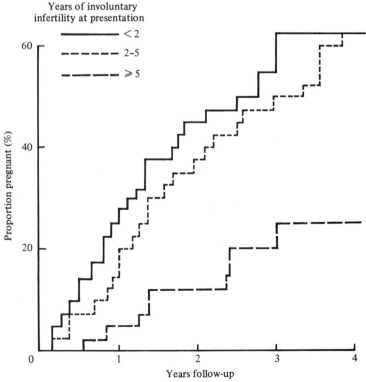

the menstrual cycle. An ovulatory menstrual cycle requires the satisfactory functioning of the pituitary–ovarian endocrine axis (which is discussed at length in Book 3, Second Edition).

If menstrual cycles fail to begin after puberty, the condition is described as primary amenorrhoea. Amenorrhoea may also be secondary, occurring after years of normal cycles, e.g. as a result of extreme starvation, as in the 'slimmers's disease' anorexia nervosa. Prolonged, strenuous exercise may also affect the menstrual cycles of healthy women (see also Book 4, Chapter 4, Second Edition). Menarche is delayed in girls who attend professional dancing schools, and menstrual dysfunction is often observed in athletes. Occasional cycles may be missed because of emotional stress, an aborted pregnancy, or discontinuation of oral contraceptive therapy.

Ovulation is the central event of the menstrual cycle, releasing the oocyte for potential conception. Approximately one in ten menstrual cycles in adult women are anovulatory, but many more just after menarche or just before the menopause (Fig. 5.4). Follicles might fail to develop because FSH is deficient, the LH surge may fail to occur or be inadequate, or the follicle may simply fail to rupture so that the oocyte remains trapped inside it. Regular menstrual cycles are not proof of ovulation, and without ovulation, conception is impossible.

Disorders of ovulation

Amenorrhoea in the presence of high blood levels of pituitary gonado-trophins (FSH, LH) indicates ovarian failure. This situation occurs

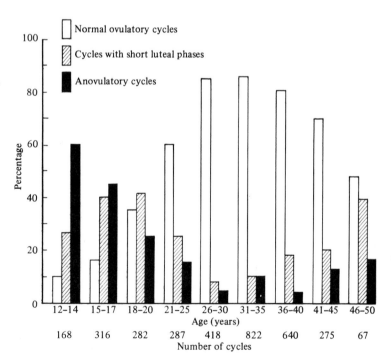

Fig. 5.4. The incidence of normal, incomplete or anovulatory menstrual cycles during a woman's reproductive life. (From R. G. Edwards. *Conception in the Human Female*. Academic Press; London (1980).)

naturally when a woman reaches the menopause; the ovary is depleted of follicles, follicular oestrogens decline and the absence of feedback control on the hypothalamus and pituitary results in an unchecked rise in FSH and LH to permanently high levels. If the woman is not menopausal, it is essential to find out if the ovary is totally or partially inactive by using various tests. A simple assay of plasma oestradiol-17β can be very informative, high levels indicating follicular activity, but care is needed to avoid confusing this with the high levels of oestrone that may be produced in peripheral fat from adrenal precursors after the menopause. If oestrogens are being secreted, injections of progesterone can be given to induce the growth of the uterine endometrium, and withdrawal bleeding ('menstruation') occurs when these injections are stopped. A positive test involves withdrawal bleeding, which demonstrates not only some ovarian activity, but also a functional uterus. A negative progesterone withdrawal test indicates complete ovarian or uterine failure. Women in this group include those with genetic disorders (e.g. XO, Turner's syndrome – see Book 2, Chapter 3, Second Edition), congenital absence of follicles, destroyed or damaged follicles, or polycystic ovaries. Of these, only the polycystic ovary syndrome can be treated by surgery and anti-oestrogens (see below).

Amenorrhoea with low levels of FSH and LH indicates a hypothalamic or pituitary failure, or a disorder of the feedback system. A lesion or tumour in the hypothalamus or pituitary is rarely found, but hyperprolactinaemia – an increased production of prolactin by the pituitary – is fairly common and often associated with galactorrhoea, where the breasts produce milk inappropriately i.e. without a preceding pregnancy. Various drugs will reduce prolactin secretion, the most widely used being the dopamine agonist bromocryptine, which also suppresses the galactorrhoea. When prolactin levels decline to normal, the menstrual cycle is usually restored.

When a woman is menstruating regularly, a number of tests can be used to see if she is ovulating:

Luteal progesterone. If ovulation occurs, a rise in progesterone secretion usually begins about 1 day after ovulation, to reach a peak at 5 days, a most reliable indicator of ovulation. The secretion of progesterone is usually accompanied by an approximate rise of 0.5 °C in basal body temperature (see Book 3, Chapter 6, Second Edition).

Endometrial cytology. The cytology of the uterine endometrium changes during the menstrual cycle. During the proliferative phase, there is an increase in cell division and a great increase in thickness of the endometrium under the influence of follicular oestrogens. The production of new cells exceeds the rate of loss, which is the opposite in the luteal phase. These histological changes in the uterus provide the basis of menstrual dating by uterine biopsies. There is a close relationship between levels of progesterone and the endometrial biopsy, but progesterone assay is preferred since it is simpler to perform and more accurate for assessing ovulation and the degree of luteal phase function.

Cervical mucus. Cervical mucus is produced under the influence of oestrogens and is maximal at the time of ovulation when it is thin, clear and acellular and conducive to sperm penetration (see also Book 3, Chapter 6, Second Edition). It changes when oestrogen decreases, becoming scant, viscous and impenetrable to spermatozoa. More than 90 per cent of the content is water, the proportion depending on the phase of the cycle. The mucus consists of glycoproteins composed of long polypeptide chains and oligosaccharide side-chains which crosslink to form multichannelled micelles, giving the mucus a specific structure: sperm penetration is possible when the channels are open. Changes in one or several constituents have been proposed to explain the variation in the physical characteristics of mucus and its interaction with spermatozoa, but a change in the quantity of water seems to be the most likely explanation. Cervical mucus is readily examined and gives an indication of the extent of oestrogen secretion in the cycle.

Treatment of disorders of ovulation

All treatments are aimed at stimulating the ovary to produce a follicle that will ovulate either naturally or in response to an injection of human chronic gonadotrophin (hCG). Ovulation must be closely synchronized with insemination to achieve fertilization. The problem with most follicle stimulation regimes is that often more than one follicle ovulates, resulting in multiple pregnancy. This leads to an increased abortion rate and an increased obstetric risk for the mother, both undesirable side-effects for an infertility treatment!

If some residual activity remains in the pituitary–ovarian axis, drugs with anti-oestrogenic activity will stimulate the discharge of gonadotrophins, presumably by inhibiting the normal negative feedback effect of gonadal oestrogen on the brain. The most widely used compound is clomiphene, which stimulates the production of pituitary gonadotrophins and hence follicular growth when given early in the follicular phase of the cycle. Oestrogen produced from the follicle might trigger an LH surge, or ovulation can be induced by hCG (Fig. 5.5).

If clomiphene treatment is unsuccessful, the ovary may respond to stimulation by exogenous gonadotrophins (Fig. 5.6). Human menopausal gonadotrophin (hMG), prepared from the urine of menopausal women, is the most widely used preparation. The treatment is fairly powerful and often recruits several follicles into growth, and careful assays for levels of oestrogens, or visualization by ultrasound scanning, can help in assessing the number of growing follicles and in avoiding hyperstimulation and superovulation.

Recently, a new method of treating ovulatory failure has been successfully applied without some of the deleterious side-effects previously mentioned. A pulsatile pump attached to the patient continuously injects gonadotrophin-releasing hormone (GnRH) at a physiological dosage, mimicking the normal rate of pulsatile discharge. This releases pituitary gonadotrophins, again in a physiological amount and pattern, which stimulate the growth of a single follicle; in virtually all cases a spontaneous LH surge then induces ovulation.

In general, the use of ovulation induction has been very successful, and no increase in congenital abnormalities has been detected among the many babies born, despite the high incidence of multiple pregnancy. However, the abortion rate in the first trimester is high, possibly because of multiple implantation or disturbances of the endocrine balance. Insufficiency of the

Fig. 5.5. Typical treatment with clomiphene to stimulate follicular growth and regulate the menstrual cycle. hCG is given routinely, although the patient may have an endogenous LH surge.

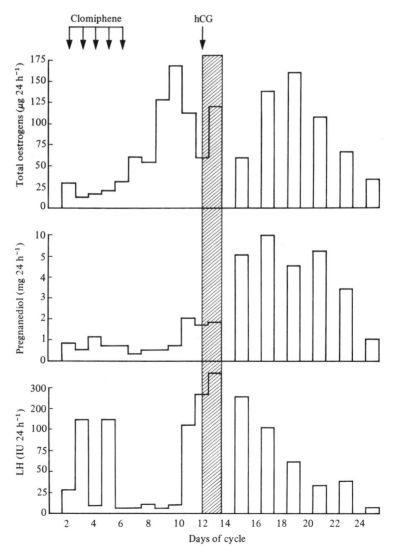

luteal phase is uncommon in women who have not been artificially stimulated, but it can be an apparent cause of infertility. The luteal phase may be too short for implantation to occur, or it may be of adequate length but associated with insufficient progesterone secretion to prepare the uterus or to maintain the pregnancy until the end of the first trimester, when the

Fig. 5.6. Follicular stimulation with human menopausal gonadotrophin (hMG), followed by hCG to induce ovulation; two cases.

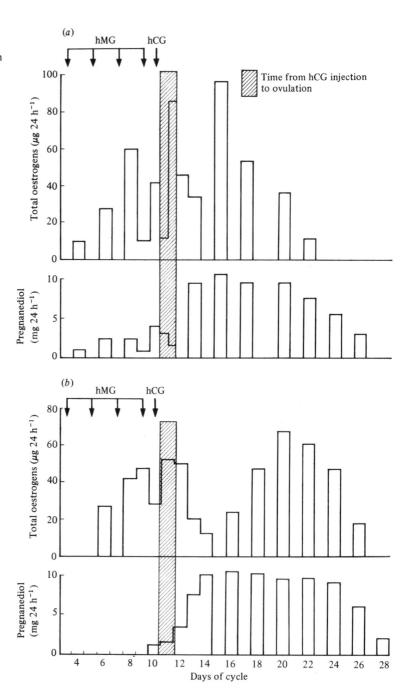

placenta takes over support. A defective luteal phase is difficult to prevent, but exogenous progesterone is effective treatment in many cases.

Having established that ovulation does occur, the next question to be decided is whether the oviduct is functional. The oviduct has a crucial and complex role in establishing a pregnancy, but investigations of its functions in infertility are largely limited to a decision about its patency.

The major cause of tubal occlusion is pelvic inflammatory disease (PID). A source of infection usually exists within the reproductive tract itself (e.g. *Neisseria gonorrhoeae*, *Chlamydia trachomatosis*), the organisms having been introduced by intercourse, an intra-uterine device, induced abortion, dilation and curettage of the uterus, etc., but inflammation can also be caused by tubal infection resulting from peritonitis or appendicitis. If PID is diagnosed and treated rapidly, the reproductive potential of the woman may not be impaired. Once damage has occurred, and adhesions have formed involving the pelvic organs, or the oviducts have become blocked, the ensuing infertility may be very difficult to treat by surgery or other means. A history of salpingitis (inflammation of the oviduct) is a common cause of ectopic pregnancy where the embryo may be forced to implant in the Fallopian tube, with life-threatening haemorrhage as a consequence. This may make it necessary to remove all or part of the tube. Complete or partial tubal occlusion can be diagnosed by two tests, hysterosalpingo-graphy and laparoscopy.

Hysterosalpingography. This is a technique in which radio-opaque material is introduced into the uterus by trans-cervical catheterization, and its passage (or non-passage) through the oviducts is viewed simultaneously by X-ray (Fig. 5.7). The procedure does not require anaesthesia, but it may be painful, and repeated or prolonged exposure of the ovaries to X-rays is not to be recommended. Unreliable results are not uncommon, and may be due to perforation of the tubes by the radio-opaque material or leakage around the cannula. Sensitivity to radio-opaque medium is sometimes observed.

Laparoscopy. Generally the final test in an infertility investigation, this procedure enables the gynaecologist actually to view the condition of the pelvic organs through an endoscope inserted through the lower abdominal wall (Fig. 5.8). During the examination, performed under general anaes-thesia, the passage of dye or air is used to demonstrate tubal patency. Simple adhesions can be freed during laparoscopy with the aid of other instruments, to allow fimbrial access to the ovary, for example. Complete or partial tubal blockage can sometimes be corrected surgically, by excising the blocked piece of tube and re-anastomosing the sectioned ends. The procedure is not simple, and microsurgery is often needed. The success rate is variable (7–24 per cent), depending on the site of the blockage, being most effective with isthmic occlusion. The oviduct may become re-occluded after surgery, and its function may be impaired, with the result that the subsequent incidence of tubal ectopic pregnancy (13 per cent) is ten times higher than normal.

Disorders of the cervix and uterus

Mucus produced by the cervix and released into the cervical canal forms a protective barrier between the relatively 'unclean' vagina and the sterile upper reproductive tract. It also functions as a 'reservoir' for motile spermatozoa and as a barrier to abnormal spermatozoa. Sperm capacitation (see Book 1, Chapter 6, Second Edition) may begin on contact with cervical mucus. Cervical disorders may be anatomical, like cervical stenosis (narrowing), or take the form of infections leading to chronic cervicitis,

Fig. 5.7. Hysterosalpingogram of a normal uterus with patent oviducts. The uterine cavity (1) is clearly outlined before the radio-opaque dye passes along the oviduct (2) and spills into the peritoneal cavity (3). (4) and (5) are shadows cast by vaginal speculum and trans-cervical catheter respectively. (From D. H. Lees and A. Singer. *Gynaecological Surgery*, vol. 5, *Infertility Surgery*. Wolfe Medical Publications; London (1981).)

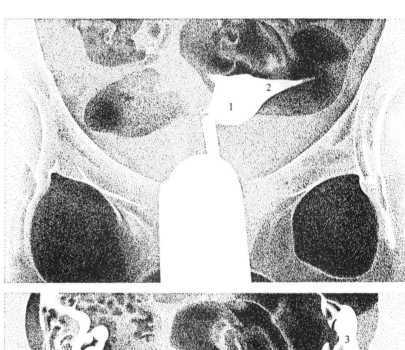

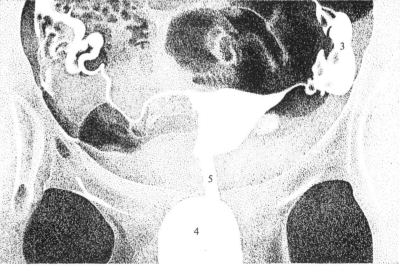

or the mucus itself may be hostile (toxic) to spermatozoa. This may be a symptom of an underlying endocrine disorder, such as low oestrogens, and so may respond to appropriate hormone therapy. If cervical hostility is not endocrinological in origin or associated with immunity to spermatozoa (see later), intra-uterine insemination with a seminal-plasma-free suspension of spermatozoa from the partner will sometimes result in conception.

Cervical function can be assessed by a post-coital test. Shortly after copulation, a mucus sample is examined for the presence of active spermatozoa. The test is not always reliable since motile spermatozoa have been found in the oviducts of women who have had a negative post-coital test. *In vitro* cervical-mucus-penetration assays can help to distinguish between a male or a female factor by testing the husband's spermatozoa with wife's and donor's mucus, and the wife's mucus with husband's and donor's spermatozoa.

The uterus is the site of implantation and growth of the conceptus, and malformations or obstructions can cause infertility. These can sometimes be simply corrected by surgery. A fairly common but difficult problem is endometriosis, the growth of endometrial tissue (the lining of the uterus) at inappropriate sites including the ovary, oviducts, bladder and peritoneal cavity. The condition can arise at any time, usually in women with patent oviducts, and it is possibly caused by retrograde menstruation, when menstrual products and fragments of endometrium pass through the oviduct and become established in abnormal sites. Apart from its deleterious effect on fertility, endometriosis can be a painful condition, as the ectopic endometrial tissue passes through the same changes as the uterine

Fig. 5.8. During laparoscopy, the organs in the pelvis are examined with the aid of a laparoscope, which is inserted through the lower abdominal wall close to the umbilicus. The abdomen is tilted slightly and inflated with inert gas to give an optimal view of the reproductive tract.

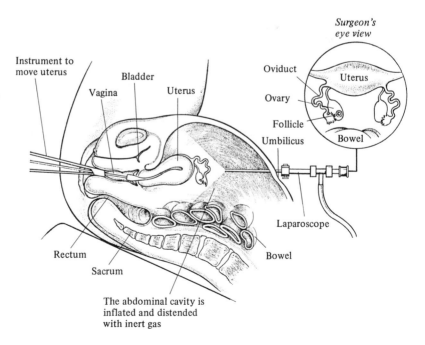

endometrium during the menstrual cycle. Treatment is usually by hormonal therapy, to depress all ovarian activity for a time in the hope that the ectopic material will regress, and this is sometimes supported by surgical removal of the plaques.

Immunological infertility

Spermatozoa are highly antigenic, and many infertile women have been shown to have high titres of plasma antibodies against spermatozoa (see also Book 4, Chapter 6, Second Edition). The antibodies may also be present in the reproductive tract, notably in cervical mucus, although these may belong to a different class of immunoglobulins. Spermatozoa can be immobilized or agglutinated, preventing fertilization. Systemic cortico-steroids may lower the antibody levels sufficiently to allow a pregnancy to be established.

Auto-antibodies to the zona pellucida surrounding the oocyte can arise in women, and this acellular investment is known to act as an antigen in experimental situations. Circulating antibodies to the zona pellucida are found in some infertile women with normal menstruation, and probably interfere with fertilization or implantation of the embryo.

The most notable recent advance in the alleviation of infertility must surely be the introduction of *in vitro* fertilization and embryo transfer (IVF and ET). Several forms of female infertility failing to respond to more conservative treatments may be overcome in this way, such as endometriosis, defective oocyte pick-up, immunity to spermatozoa and tubal occlusion.

Male infertility

Twenty-five years ago, andrology, the study of male infertility, scarcely existed as a medical speciality. Now, thanks to more enlightened social attitudes and medical progress, infertility is seen as a problem for the 'couple'. Knowledge of male reproduction is improving constantly, although treatments for male infertility are still largely unsuccessful.

The diagnosis of male infertility is not straightforward. Objective tests to assess the fertilizing capacity of spermatozoa from supposedly infertile men have only recently become available by the use of *in vitro* fertilization of human or animal oocytes. These tests are available at only a few infertility centres, and they are expensive, laborious, difficult to interpret, and not applicable to all cases. The vast majority of infertility investigations evaluate the male partner on the basis of a simple semen analysis.

Semen analysis

Semen is a suspension of spermatozoa (produced by the testes) in the secretions from the male accessory glands, mainly (in man) the prostate and the seminal vesicles. Biochemical analysis of semen for fructose and acid phosphatase can establish the function of these glands, but provides no indication of sperm fertilizing capacity. A microscopic study of semen gives a quantitative analysis of sperm concentration, motility and morphology (Table 5.1), but it is not a test of fertility; a 'normal' result does not preclude male infertility. Several causes of infertility can nevertheless be identified. If no spermatozoa are present (azoospermia), conception is impossible, but the threshold number is difficult to assess, though commonly taken as 5×10^6 ml^{-1}. Similarly, if only immotile spermatozoa are present, no matter how high the concentration, conception is impossible, since sperm motility is essential for the penetration of the cervical mucus and the investments of the oocyte. Every ejaculate contains some abnormally formed spermatozoa (which can be motile or immotile), but absolute criteria for sperm normality have yet to be established, and men with virtually no 'normal' spermatozoa can still father children. However, homogeneous morphological abnormalities, such as lack of the acrosome or decapitation of the spermatozoa, are irreversible and untreatable causes of male sterility. (For more detailed information, see the WHO manual edited by M. A. Belsey *et al.* listed in Suggested further reading.)

In vitro *tests of spermatozoa*
The ideal system for testing the fertilizing capacity of human spermatozoa is clearly human *in vitro* fertilization (see later). Second best at present is the hamster oocyte test, in which zona-free hamster eggs are incubated with specially prepared (seminal-plasma-free) human spermatozoa. Capacitated spermatozoa (incubated at 37 °C in a seminal-plasma-free medium) readily bind to the egg membrane and some penetrate the cytoplasm and form pronuclei. The percentage of eggs penetrated often gives a fairly reliable measure of fertilizing ability. Samples from fertile men score significantly better in this test than those from infertile men. The hamster egg test so far provides the only test for spermatozoa that correlates with the duration of the infertility (Fig. 5.9).

Table 5.1. *Semen analysis: various parameters examined routinely in an infertility investigation*

	Possible range	Normal range
Sperm concentration ($\times 10^6$/ml)	0–500	10–250
Motile (%)	0–80	> 20
Living (%)	0–90	40–90
'Normal' morphology (%)	0–70	20–70
Linear progression*	1–4	3–4
Agglutination*	1–4	< 2
Viscosity* } { measured after 2 h	Low–high	Low
Liquefaction* }	Incomplete–complete	Complete
Spermatogenic cells	Not present–present	
Inflammatory cells	Not present–present	

* These parameters are scored subjectively.

Causes of male infertility
As with female infertility, the causes of male infertility are as complex as male reproduction itself.

Endocrinology. Spermatogenesis takes place within the seminiferous tubules, and the process is dependent on a high intratesticular androgen concentration (see Book 1, Chapter 4, and Book 3, Chapter 4, Second Edition) produced by the Leydig cells. The entire process depends on FSH and LH, and perhaps prolactin, which are secreted by the pituitary. Primary endocrine failure (hypopituitary hypogonadism) is a rare cause of infertility, but it responds to gonadotrophin therapy. High FSH levels are often a sign of damage to the testes. FSH is believed to be under the negative feedback control of the seminiferous tubule hormone inhibin, and conditions where the seminiferous epithelium is damaged will result in reduction of inhibin and elevated FSH.

Testicular defects. The development of the seminiferous tubules starts at an early age and spermatogenesis may be established by the age of 11. In cases of undescended testes (unilateral or bilateral), the corrective surgery

Fig. 5.9. Correlation between the duration of infertility (in years) and hamster egg test results (expressed as the percentage of zona-free eggs penetrated by human spermatozoa). (From Cohen *et al.*, *Fertil. Steril.* **37**, 565 (1982).)

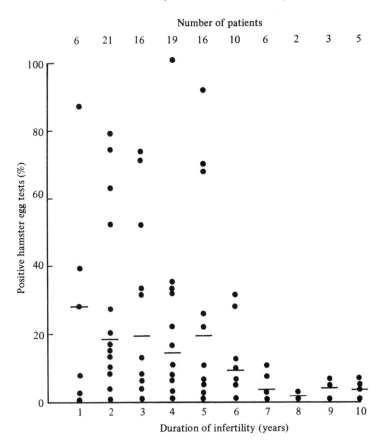

must be performed before the age of 6 years to prevent destruction of the tubules at the temperatures found within the abdomen. The testes in the scrotum are normally about 2.5 °C cooler than basal body temperature, and this temperature differential is crucial to successful spermatogenesis in man.

Intrascrotal temperature can be raised by a varicocele (dilatation of the spermatic vein), and could cause the subfertility sometimes associated with this condition (Fig. 5.10). Varicocele arises through inadequate venous drainage from the internal spermatic vein, and almost invariably occurs on the left side; it is common in the general population (10–15 per cent) but has a higher frequency among the infertile. The mechanism whereby it affects testicular function, if at all, is unclear; theories range from increased intrascrotal temperature, to retrograde flow of substances from the adrenal gland down the spermatic vein, or possibly hypoxia of the germinal epithelium. Surgery, by division and ligation of the internal spermatic vein, is regularly used to correct varicoceles, but its therapeutic effects on infertility are debatable.

In men with a supernumerary X-chromosome (XXY or XXXY), spermatogonia degenerate prematurely so that effective spermatogenesis is absent (Klinefelter's syndrome – see also Book 2, Chapter 3, Second Edition). In germ cell aplasia the germinal epithelium is composed entirely of Sertoli cells, possibly arising through a failure of the germ cells to migrate to the primitive gonad.

In other men the tubules are damaged secondarily, e.g. by mumps virus after puberty or by other diseases. Many chemicals are now considered hazardous to fertility, although it is often difficult to prove the cause conclusively. Organic chemicals, such as propanolol, ethanol, nitrofurans, and heavy metals (lead, cadmium, mercury), as well as some drugs and pesticides have each been suspected to impair male fertility. DBCP (dibromochloropropane), a toxin used to kill nematode parasites, can result in testicular damage, and the devastating effects on fertility due to

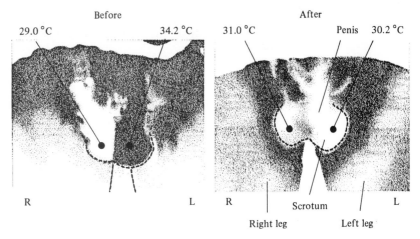

Fig. 5.10. Effect of varicocelectomy on scrotal temperature in an infertile man – a frontal view by thermography. After surgery, left side is considerably cooler. (From W. F. Hendry. Thermography. In *Male Infertility*. Ed. T. B. Hargreave. Springer-Verlag; Berlin (1983).)

cancer chemotherapy with antimitotic agents such as cyclophosphamide, methotrexate, chlorambucil, melphalan, and nitrogen mustard are well documented. Male sterility (azoospermia) was encountered 25 years ago in China when workers preparing cottonseed oil were suddenly exposed to high concentrations of a yellow cottonseed pigment called gossypol, due to a change in the processing technology. The drug has since been tested as a male contraceptive in clinical trials in China, but the results so far are not promising (see Chapter 2).

Men may be infertile due to obstruction of the vas deferens or epididymis, which is usually congenital. In these cases the semen does not contain any spermatozoa. Corrective surgery (epididymo-vasostomy, to join the vas to the epididymis above the site of obstruction) can be performed, but the success rate is not high.

Autoimmunity. Spermatozoa are not present in the fetus when the immune system develops, hence immunological tolerance to sperm-specific antigens may never be established. This lack of tolerance is unimportant so long as spermatozoa do not come into contact with the immune system. In cases of testicular injury or disease, spermatozoa can escape into surrounding tissues, and autoantibodies can be formed which cause the agglutination or immobilization of ejaculated spermatozoa (see also Book 4, Chapter 6, Second Edition). The severity of the condition is related to the titres of antibody, and different immunoglobulins, IgG, IgM, IgA, might be present in blood and seminal plasma. Fertility can sometimes be restored after corticosteroid therapy to reduce the antibody titre.

Impotence. Erectile or ejaculatory failure can be psychogenic, neurological or vascular in origin (see Book 8, Chapter 2, First Edition). Patients with irreversible erectile impotence may be helped by Giles Brindley's recent discovery that injection of a smooth muscle relaxant, such as papaverine or guanethidine, into the corpora cavernosa of the penis can cause an immediate erection that will last for hours, depending on the dose injected. Some patients may benefit from the insertion of a penile implant into the corpora cavernosa; they are available in a wide range: rigid, flexible or inflatable.

One kind of ejaculatory failure is retrograde ejaculation, where the semen is forced back into the bladder rather than out through the urethra, owing to obstruction or bladder-neck incompetence. About 50 per cent of men with this problem can ejaculate normally with a full bladder and, if this fails, fertile spermatozoa can sometimes be retrieved from the bladder after altering the pH and osmolarity of the urine by oral doses of bicarbonate plus large amounts of water.

Other problems. Even after excluding all the above-mentioned causes, infertility in the majority of subfertile men is still unexplained. The largest group includes those with some motile spermatozoa in the ejaculate and an apparently normal partner.

Over the last 20 years, many different non-specific treatments for such infertility have been proposed, but almost all are only weakly effective (Table 5.2). After repeated failures to improve male infertility, many couples accept artificial insemination with donor semen or adoption. The law concerning the rights of children conceived in this way (and indeed, the rights of the donor!) are constantly being questioned in many countries, but it remains a widely accepted practice. Perhaps the increasing success of human *in vitro* fertilization as a means of treating both male and female infertility will make donor insemination less popular.

In vitro fertilization

Since the pioneering work by Patrick Steptoe and one of us (RGE) produced Louise Brown, born in 1978, the world's first child conceived outside its mother's body, human *in vitro* fertilization (IVF) has been both applauded and condemned from all sides. The subject is highly emotive and charged with difficult ethical issues, so it is not surprising that it surpasses even heart transplants for media appeal. While discussions go on, the number of so called 'test-tube babies' (possibly over 20000 by now) and delighted parents continues to increase, and IVF has become established as a household phrase. Most couples requesting IVF have had a long period of infertility, and conventional treatments have been tried and failed; IVF is usually their last hope.

Although most obvious as a method of achieving conception for women with tubal defects, many types of both male and female infertility have been successfully circumvented by IVF – absent or occluded oviducts, ovaries

Table 5.2. *Non-specific procedures and treatments for subfertile men*

A.	To improve spermatogenesis	
	clomiphene	corticosteroids
	gonadotrophins	thyroxine
	androgens	psychotropic drugs
	'rebound therapy'	vitamins
	bromocriptine	minerals
B.	To improve sperm motility (*in vitro*)*	
	caffeine	
	carnitine	
	kinins	
C.	To improve sperm quality (*in vitro*)*	
	split ejaculate collection	glass wool columns
	proteolytic enzymes	sephadex columns
	donor seminal plasma	sperm migration into buffer

* These treatments are followed by artificial insemination of the wife with the treated sperm preparation.

enclosed in adhesions, defective oocyte pick-up, endometriosis, male or female immunity to spermatozoa, oligospermia and unexplained (idiopathic) infertility (Table 5.3). Many couples suffer from more than one problem. In all cases, the basic procedure is the same:

> oocyte recovery from the ovarian follicles,
> insemination *in vitro*,
> embryo cleavage *in vitro* and subsequent replacement in the uterus, with or without cryopreservation.

Oocyte recovery

Today, untreated menstrual cycles are rarely used for oocyte recovery; virtually all patients are stimulated with clomiphene or gonadotrophins. This increases the probability of several follicles developing, and hence of several oocytes being collected for attempted fertilization. Follicle development is monitored indirectly by measuring the daily urinary output of oestrogen, or directly by ultrasound scanning of the ovaries to measure follicular size (Fig. 5.11). The timing of impending ovulation (and therefore oocyte collection which must *precede* it) is achieved either by monitoring LH secretion to detect the onset of the endogenous pre-

Table 5.3. *Bourn Hall Clinic experience, 1983–84*

Overall data
1005 laparoscopies for oocyte recovery performed

969 oocytes recovered	96% of laparoscopies
726 patients had at least one embryo replaced	73% of laparoscopies
189 normal pregnancies were established	19% of laparoscopies
78% of oocytes inseminated were fertilized	

Success of IVF for different causes of infertility (incidence of birth per oocyte recovery)*

Tubal infertility and frozen donor semen (husband azoospermic)	25%
Cervical mucus hostility	24%
Tubal infertility – reproductively normal male partner	19%
Male infertility – reproductively normal female partner	17%
Unexplained infertility	14%
Both partners infertile	13%

Incidence of pregnancy per embryo replacement

One embryo replaced	15.9%	pregnant
Two embryos replaced	25.3%	pregnant
Three embryos replaced	35.5%	pregnant

Incidence of abortion

Patients aged		
	20–24 years	0%
	25–29 years	20%
	30–34 years	30%
	35–39 years	29%
	⩾ 40 years	53%

Only two successful pregnancies with women over age 41
Ectopics (tubal) 5

---

* Data based on the first 8 months of 1984.

ovulatory LH surge, or by injecting hCG to induce ovulation when the follicles reach optimal size. Oocyte recovery is scheduled when the oocytes should be mature (ripe), a few hours before the predicted time of ovulation. Laparoscopic oocyte recovery requires general anaesthesia and peritoneal insufflation, whereas oocyte recovery by ultrasound scanning requires only local anaesthesia and distension of the bladder with saline solution (Fig. 5.11). The contents of the pre-ovulatory follicle are aspirated through a

Fig. 5.11. Ultrasound scan of the lower abdomen which renders visible the follicles in the ovaries. The large black area at the top of the picture is the bladder. The ovary on the left has several clear follicles; the other is not visible in this scan.

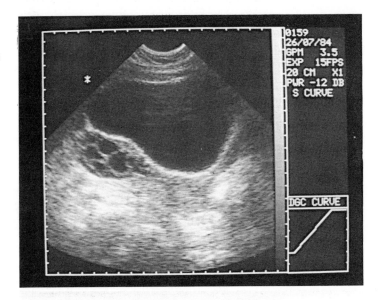

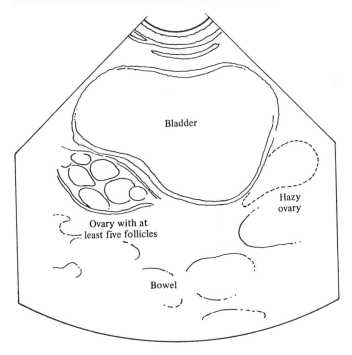

long needle into a collecting pot (Fig. 5.12). The aspirate is then examined to find the oocyte in its cumulus mass, and if unsuccessful the follicle is flushed with heparinized culture medium and reaspirated. At present, laparoscopic oocyte recovery has a higher success rate that ultrasound oocyte recovery.

The oocytes are placed in a relatively simple bicarbonate-buffered culture medium supplemented with maternal serum and sometimes follicular fluid. Unripe oocytes are sometimes collected, and maturation may sometimes be achieved by culture overnight with follicular fluid from a ripe follicle. Culture takes place in droplets under medicinal paraffin in Petri dishes, or in culture tubes, keeping the medium constant with respect to temperature (37 °C), pH and osmolarity. Semination time is highly

Fig. 5.12. Apparatus used for collecting the contents of the follicle, including the oocyte.

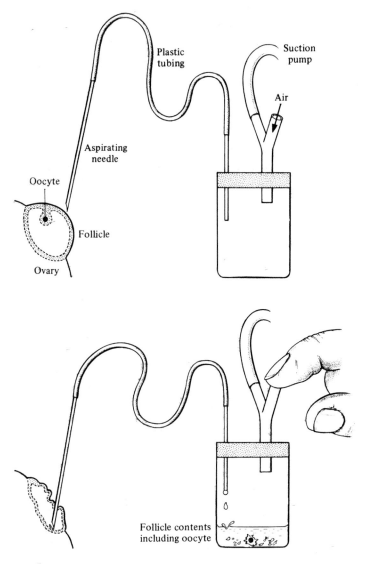

variable (0–24 h after oocyte recovery), depending on the ripeness of the oocyte, but 6 hours is approximately average.

Semination of oocytes

Most of the semen used in IVF is produced by masturbation at about the time of oocyte recovery, although stored frozen semen has also been used successfully. Seminal plasma inhibits fertilization, so semen preparation basically consists of producing a seminal-plasma-free stock suspension of predominantly motile spermatozoa. Normal semen is readily prepared by simple washing involving centrifugation with artificial medium, but the procedure may be considerably more complex in cases with poor semen, from which it may be necessary to remove immotile spermatozoa, phagocytic cells, debris or sperm clumps, or concentrate all the available motile spermatozoa. Prepared stock suspensions of spermatozoa are normally stored at room temperature for a few hours to permit sperm capacitation to occur, then the oocytes are placed in droplets of medium containing 50 000–100 000 motile spermatozoa/ml for fertilization. On the next day, 'Day 1', the oocyte may be examined for the presence of two pronuclei, evidence of successful fertilization. Fertilization rates are usually in excess of 75 per cent for ripe oocytes.

Embryo replacement

After fertilization has taken place, the embryo is cultured for 1–2 days to assess normal cleavage and development. On Day 2, the embryo has approximately four cells, and the majority of embryos are replaced in the uterus between this stage and the eight-cell stage, which is found on Day 3 (Fig. 5.13). Pregnancies have also been achieved by replacing zygotes (Day 1) or blastocysts (Days 5 or 6). Embryo replacement is usually a non-traumatic, straightforward process requiring no anaesthesia, during which the embryo is deposited in the uterus via a narrow, flexible plastic catheter introduced through the cervix. The chance of pregnancy is increased by increasing the number of embryos replaced (Table 5.3), hence the preference for more than one oocyte, but the total number of embryos replaced depends on the policy of individual clinics. Some clinics replace only three to avoid multiple pregnancies, while other clinics find it necessary to replace as many as six or seven to have a reasonable chance of one implanting. The desire to establish a pregnancy by replacing all available embryos must be balanced against the increased abortion rate and high obstetric risk to the mother in multiple pregnancies, to say nothing of the stress associated with bringing up triplets or quadruplets.

Now that human embryos can be frozen and thawed successfully, thanks to the work of Alan Trounson and Linda Mohr in Melbourne, the number of 'fresh' embryos replaced can be kept to a minimum – the optimum could be two – and the remainder frozen for storage until required for replacement on a subsequent natural (untreated) cycle. Cryopreservation and storage of embryos has been routine practice for many years with embryos from laboratory or domestic animals, notably cattle. The embryos are protected from the damaging effects of ice formation by the addition of a cryoprotectant substance, either glycerol or dimethylsulphoxide (DMSO). The temperature is lowered slowly to between $-16\,°C$ and

Fig. 5.13. Human embryos; both the four-cell (one of the cells is out of focus) (*a*) and the eight-cell embryos (*b*) have spermatozoa attached to the outside of the zona pellucida. The majority of human embryo replacements take place at about these stages.

(*a*)

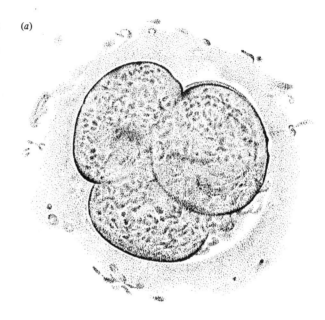

(*b*)

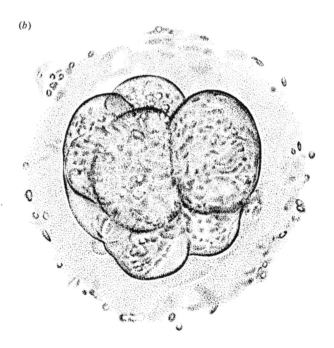

−80 °C, before plunging the embryos into liquid nitrogen (−196 °C) for indefinite storage. Cryopreservation may also solve the problem of 'spare' embryos, not replaced in their mother but available for media testing or for other investigations designed to improve methods of treatment. This is, however, a procedure that raises many ethical objections.

Based on the experience of the past 2 years, on average about 30 per cent of women having embryos replaced become pregnant. The figure is influenced by various factors, the age of the woman and the number of embryos replaced being the most significant (Table 5.3). This success rate may improve as IVF is still relatively new, but it is already comparable to the success rate achieved by most conventional forms of infertility therapy. It must also be stressed that the majority of IVF patients had previously been considered beyond help, since all conventional therapies had been tried without success, and they were therefore of an age when natural fertility is already on the wane.

Future prospects

The incidence of infertility in some societies could be reduced in the future by preventive rather than curative measures. In societies where contraception is not freely available, abortions (often multiple) become the only way of avoiding children, with all the attendant risks of damage and infection of the female reproductive tract that can end in infertility and even death (see Chapters 3 and 4). This is especially true in cases of 'backstreet' abortion. Similarly, in societies where venereal disease is prevalent (especially gonorrhoea), both male and female fertility are adversely affected, usually owing to the occlusion of the oviducts or vas deferens. Infertility that is 'self-inflicted' in this way should be totally avoidable with good health education and medical care.

Follicle stimulation is already becoming more controllable and refined by the use of 24-h pulsatile administration of GnRH, to produce a reproductive cycle close to the physiological norm. The patient wears a small portable pump connected to a fine needle under the skin below the umbilicus. The GnRH is injected in tiny amounts delivered every few minutes until a follicle has grown and ovulation has occurred. Treatments like this should replace the earlier more 'blunderbuss' methods of stimulation. It may even be possible to find some therapy to stimulate spermatogenesis in oligospermic men, or perhaps to improve the ratio of motile to immotile spermatozoa.

The biggest recent advance in infertility treatment – IVF – will no doubt be one of the procedures of choice for the future. It should now be considered before more invasive methods like radical tubal surgery or varicocelectomy, which have possible deleterious side-effects. If the woman's pelvic organs have too many adhesions to allow even laparoscopy, oocytes can be recovered with the aid of ultrasound scanning without general anaesthesia.

In vitro fertilization and embryo transfer have many possibilities not yet exploited, partly because society has had trouble deciding what is ethically acceptable practice. An enquiry committee was, however, set up in 1982 by the UK Department of Health and Social Security and its final report, usually referred to as the Warnock Report after its chairman Dame Mary Warnock, was published in 1984. The enquiry examined the ethical, social and legal implications of new and potential developments in the field of human fertilization and embryology, and on this basis the committee made 63 recommendations. In the committee's view, the services of donor insemination, IVF, egg donation and frozen embryo storage should continue to be available, but a statutory licensing authority should be established to regulate the conduct of research and infertility services. The committee accepted the use of only one form of embryo donation, that in which a donated egg was fertilized *in vitro* with the husband's spermatozoa and not where the embryo was recovered from another woman. Definite legal limits were proposed for research on embryos, and any *un*authorized use of embryos *in vitro* up to day 14 after fertilization would be considered a criminal offence; *all* research on living embryos beyond day 14 would be considered a criminal offence.

Donor spermatozoa have already been routinely used for IVF with couples where the wife has tubal blockage and the husband is azoospermic, but the use of donor oocytes, for couples where the husband is fertile but the wife has no ovaries or has untreatable ovarian dysfunction, has not been widely introduced. Embryo donation from one couple to another, basically pre-implantation adoption, was, as we have seen, unfavourably regarded by the Warnock Committee, but stronger opposition was mounted against the organized use of surrogate mothers – women willing (sometimes for a considerable fee, sometimes without payment) to carry a baby to term for a woman who is incapable of pregnancy – which the committee proposed should become a criminal offence.

Couples who have endured voluntary 'infertility', because they have a known genetic defect they do not wish to transmit and are willing to accept abortion if their fetus is diagnosed as having the defect, can possibly be helped by IVF. Their embryos could be grown to the blastocyst stage (Day 5 or 6) and some cells removed for diagnosis (see Book 2, Chapter 6, Second Edition). The procedure does not affect the embryo. If the test can be performed rapidly, as is the case with sexing by DNA probes or other methods of identifying the Y-chromosome, embryos of the desired sex could be replaced, so avoiding sex-linked disorders such as haemophilia and muscular dystrophy. If the test proved to be too slow to allow immediate replacement of the embryos – and many karotyping procedures require several days – the embryos could be frozen and only those with an acceptable chromosomal complement thawed and replaced during a later cycle.

Many manipulative procedures that have been successfully accomplished with the oocytes and embryos of other mammalian species could theoretically be applied to human embryos. It is now possible to microinject multiple copies of a foreign gene into the pronucleus of a fertilized egg, and get integration of that gene into the 'host' genome. The foreign gene replicates along with the 'host' DNA, and the foreign protein is subsequently synthesized with the 'host' proteins. This has been accomplished in mice with the rabbit globin gene. Even more dramatically, Richard Palmiter and his colleagues (see Chapter 1 and Suggested further reading) have produced mice larger than normal by injecting rat or human growth hormone genes into the pronuclei of mouse eggs. It is, of course, desirable to have the regulator gene for the foreign gene also coupled into the system, to allow biological feedback control to function. This technique could allow desirable or missing genes to be introduced into the human embryo; for example, in cases with a known risk of diabetes, the genes for the synthesis and regulation of insulin could be introduced.

Some men have too few motile spermatozoa even for IVF, but with micromanipulation techniques it should be possible to deposit a few motile spermatozoa inside the zona pellucida. If all are immotile, a single spermatozoon or a sperm head might be microinjected into the oocyte cytoplasm, though we do not know yet whether normal fertilization would follow such a procedure.

Several commercial cattle embryo companies offer to supply divided embryos capable of developing into identical twin calves. This technique has yielded very encouraging results in cattle, and could be useful for human IVF where only single oocytes are recoverable. In the sheep, work by Steen Willadsen and one of us (C.F.) has shown that it is possible to produce up to five identical progeny (quintuplets) from a single embryo

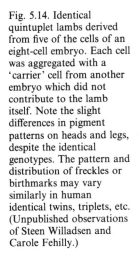

Fig. 5.14. Identical quintuplet lambs derived from five of the cells of an eight-cell embryo. Each cell was aggregated with a 'carrier' cell from another embryo which did not contribute to the lamb itself. Note the slight differences in pigment patterns on heads and legs, despite the identical genotypes. The pattern and distribution of freckles or birthmarks may vary similarly in human identical twins, triplets, etc. (Unpublished observations of Steen Willadsen and Carole Fehilly.)

(Fig. 5.14), by dissociating and reaggregating early cleaving embryos. The potential of some mammalian embryos is clearly in excess of the one fetus that is normally produced, and with embryo division, the possibility of identical twins, triplets or more 'to order' becomes very real. Such monozygotic sibs do, of course, represent a *finite* clone, as the physiology and size of the embryo limit the number of viable conceptuses that can be produced by simple division or blastomere separation, but it may eventually be possible to create much larger clones from an existing individual by removing the nucleus from a fertilized oocyte and replacing it with a nucleus from the selected individual. This has been achieved in frogs (see Fig. 6.4) but not, as yet, in mammals. The initial step of enucleating the oocyte is straightforward, but getting a suitable replacement nucleus is more difficult, as adult cells have specialized nuclei. Certain types of tumour cells (those of teratocarcinomata) behave and develop like embryonic cells, so it may be possible to identify the trigger or stimulus that reverts a specialized cell to the pluripotential state found in an early embryo. Indeed, a recent report by Bradley and his colleagues (see Suggested further reading) indicates that cells deriving from embryos can display the pluripotential faculty if cultured under specific conditions, without passing through the carcinoma phase. These cells can then contribute to chimaeric animals, normal in apparently all tissues and chimaeric even in the germ line.

Such is the rate of progress of modern technology and science that in the not-too-distant future the option to reproduce asexually may become a commonplace. Then the trials and tribulations of infertility will burden only those who choose to reproduce sexually – the old-fashioned way.

Suggested further reading

Endocrinology of female infertility. D. T. Baird. *British Medical Bulletin*, **35**, 193–8 (1979).

Formation of germ-line chimaeras from embryo-derived teratocarcinoma cell lines. A. Bradley, M. Evans, M. H. Kaufman and E. Robertson. *Nature*, **309**, 255–6 (1984).

In vitro fertilization: a treatment for male infertility. J. Cohen, R. G. Edwards, C. B. Fehilly, S. B. Fishel, J. Hewitt, J. M. Purdy, G. Rowland, P. C. Steptoe and J. M. Webster. *Fertility and Sterility*, **43**, 422–33 (1985).

Pregnancies following the replacement of cryopreserved expanding human blastocysts. J. Cohen, R. F. Simons, C. B. Fehilly, S. B. Fishel and R. G. Edwards. *Journal of* In Vitro *Fertilization and Embryo Transfer*, **2**, 59–64 (1985).

Analysis of 25 infertile patients treated consecutively by *in vitro* fertilization at Bourn Hall. S. B. Fishel, R. G. Edwards and J. M. Purdy. *Fertility and Sterility*, **42**, 191–7 (1984).

The etiology of pelvic inflammatory disease. L. Keith and G. S. Berger. *Research Frontiers in Fertility Regulation*, **3**, 1–15 (1984).

Deep freezing and transfer of human embryos. L. R. Mohr, A. O. Trounson and L. Freeman. *Journal of* In Vitro *Fertilization and Embryo Transfer*, **2**, 1–10 (1985).

Ultrasonic assessment of endometrial changes in stimulated cycles in an *in vitro* fertilization and embryo transfer programme. B. Smith, R. Porter, K. Ahuja and I. Craft. *Journal of* In Vitro *Fertilization and Embryo Transfer*, **1**, 233–8 (1984).

Observations on 767 clinical pregnancies and 500 births after human in-vitro fertilization. P. C. Steptoe, R. G. Edwards and D. E. Walters. *Human Reproduction*, **1**, 89–94 (1986).

In vitro fertilization and embryo transfer (bibliography with review). A. O. Trounson and P. Lutjen. *Bibliography of Reproduction*, **44**, B1–B6 (1984).

Microsurgery of the Fallopian tube: from fantasy to reality. R. M. L. Winston. *Fertility and Sterility*, **34**, 521–30 (1980).

Factors influencing pregnancy rates following *in vitro* fertilization and embryo transfer. C. Wood, R. McMaster, G. Rennie, A. Trounson and J. Leeton. *Fertility and Sterility*, **43**, 245–50 (1985).

Zona-free hamster eggs: their use in assessing fertilizing capacity and examining chromosomes of human spermatozoa. R. Yanagimachi. *Gamete Research*, **10**, 187–232 (1984).

Infertility, Diagnosis and Management. Ed. J. Aiman. Springer-Verlag; New York (1984).

Human Sexuality and its Problems. J. Bancroft. Churchill Livingstone; London (1983).

Fertilization of the Human Egg in Vitro. Ed. H. W. Beier and H. R. Lindner. Springer-Verlag; Berlin (1983).

Laboratory Manual for the Examination of Human Semen and Semen–Cervical Mucus Interaction. Ed. M. A. Belsey, R. Eliasson, A. J. Gallegos, K. S. Moghissi, C. A. Paulsen and M. R. N. Prasad. Press Concern; Singapore (1980).

Conception in the Human Female. R. G. Edwards. Academic Press; London (1980).

Human Conception in Vitro. Ed. R. G. Edwards and J. M. Purdy. Academic Press; London (1982).

Male Infertility. Ed. T. B. Hargreave. Springer-Verlag; Berlin (1983).

The Infertile Couple. R. J. Pepperell, B. Hudson and C. Wood. Churchill Livingstone; Edinburgh (1980).

Warnock Report. See *Report of the DHSS Committee of Inquiry into Human Fertilisation and Embryology* in Suggested further reading, at the end of the next chapter.

6

Reproductive options, present and future

ANNE McLAREN

'*The strongest principle of growth lies in human choice.*' George Eliot

Family planning, in the sense of using contraception to determine when you have children and how many you have, is now an accepted part of the way of life of most couples in the developed countries, and many couples in the developing world. If this were not so, the future outlook for our species would be grim: the crucial role that contraceptives have to play in solving the problem of over-population has been stressed earlier in this book. From the point of view of the individual, too, safe and effective methods of contraception are a boon. Few women would want to return to a situation where the only alternative to sexual abstinence was a series of pregnancies, each more unwanted than the last.

Yet there is another side to the coin. Starting a family means for most people a change in their life-style and in their image of themselves; bringing a new person into the world is a major responsibility, and if it is to be determined by choice, not chance, the decision may be an onerous one. For women, the decision is particularly hard because of the time factor. If they have an enjoyable and satisfying job that may be difficult to combine with motherhood, they will want to put off getting pregnant; yet they know that if they put if off too long, their fertility will decline and they may never have the children they want.

In the future, the term 'family planning' is going to take on a new significance. Recent advances in our understanding of reproduction, development and genetics mean that we are going to be able more and more to control the reproductive process, not only preventing unwanted pregnancies but helping infertile couples to have children, preventing the birth of abnormal babies, perhaps curing genetic diseases, and perhaps controlling the genetic characteristics, including the sex, of the babies that are born. Each of these steps lessens the amount that is left to chance and broadens the area in which individuals or couples can exercise choice and so may be faced with the responsibility of taking decisions. In this chapter I shall try to outline which types of 'family planning' in this wider sense are already possible, which may become possible within the reproductive lifetime of those who read this book, and which are likely to remain in the realm of science fiction until at least the next century. If people are to exercise more choice in planning their families, it is essential that they should have the fullest information to help them in their decisions.

Alternatives to adoption

Donor insemination

About one in ten of all couples have an infertility problem, and in at least a third of such cases the problem can be traced to the man (see Chapter 5). Sometimes, of course, treatment is effective; but when no normal spermatozoa are being produced, as a result of some previous testicular infection, or some genetic defect such as Klinefelter's syndrome, there is little that can be done to improve the situation. If the couple want a child badly enough, they may request, or be offered, artificial insemination by donor.

For some couples, adoption offers an alternative to donor insemination, but with fewer unwanted pregnancies going to term and less discrimination against unmarried mothers, the number of babies available for adoption in most developed countries is now very small. In any case, many couples would prefer to have a baby that was both genetically and physiologically the mother's own, rather than to adopt one that was unrelated to either parent. Donor insemination has been used in clinical practice for several decades, and thousands of babies have been conceived in this way in Britain alone; yet, although it has never been illegal, it has only recently come to be regarded as a socially acceptable procedure.

The success rate is high: most women conceive after a few months of treatment, and many couples return for a second child. Frozen semen, which it is thought can be preserved indefinitely at the temperature of liquid nitrogen, is now widely used, as it allows the woman to be inseminated at any time and makes it easier to select a donor sharing some of the husband's physical characteristics.

No increase in the congenital malformation rate is seen with donor insemination, whether fresh or frozen semen is used; if anything, the proportion of abnormal conceptions tends to be lower than usual, since only men in good health and with no family history of genetic defect are accepted as donors. In most cases, the identity of the donors is strictly confidential, and is never revealed to the patients; nor do the donors know to whom the semen is given, or whether any pregnancies have resulted. Very extensive use of the semen from any one donor is discouraged, to minimize the risk of spreading some unrecognized genetic defect, but clinics differ in the number of pregnancies they allow from each donor.

The use of artificial insemination in clinical practice is minute compared to its use in the cattle industry. In Britain today, three million cattle are bred by artificial insemination (AI) each year, mostly from frozen semen. A genetically superior bull could in this way sire as many as 200 000 calves annually. Genetic improvement resulting from AI has been an important factor in the great increase in milk yield, and particularly in milk fat content, seen in cattle during the last 30 years (see also Chapter 1).

People sometimes ask whether donor insemination could be used to effect genetic improvement in our own species. Controlled breeding of any sort, whether by artificial or by natural insemination, could probably produce fairly rapid changes in such physical characteristics as height or skin colour, but on the crucial question of which direction of change would constitute 'improvement', opinion might be divided. When it comes to the more elusive characteristics by which we value our fellow human beings, such as honesty and amiability, the likelihood that they could be intensified by selective breeding seems small; while as far as general health and vigour are concerned, improvements resulting from better nutrition and medical care would far outweigh anything that could be expected from genetic manipulation. Recently, a group of Nobel prize winners in the USA proposed to bank their semen so that women could volunteer to bear their children by artificial insemination. Even if any pregnancies are achieved, it will be hard to assess the results of the 'experiment'; but few biologists would be sanguine that this group of donors would prove more satisfactory than a group of, for example, medical students.

IVF and proembryo* transfer

In couples where the infertility problem can be traced to the woman, big advances have been made by medical science in the effectiveness of treatment, in particular in the hormonal control of ovulation. These are reviewed in Chapter 5.

Treatment with gonadotrophic hormones is widely used in animal breeding to increase the number of eggs shed by cows, ewes and sows of high genetic worth. The 'superovulated' eggs can all be fertilized in the normal way, but to avoid overcrowding in the uterus the proembryos are recovered before implantation and transferred to a series of foster-mothers synchronized to receive them. Proembryos can also be stored in liquid nitrogen for months or years before transfer, and in the frozen state can be transported easily and cheaply from one country to another. Pioneered in the mouse, frozen storage has now been successfully extended to many other species, including the rat, cow, sheep, goat, marmoset monkey and man (Fig. 6.1). Frozen proembryo banks allow the gene pool of rare breeds of domestic animals to be preserved, and may have an important role to play in conserving some species of wild animals that are in danger of extinction. From the point of view of animal improvement, superovulation followed by proembryo transfer is the female equivalent of AI and, like AI, is of considerable economic importance. In several countries, proembryo transfer and also proembryo freezing are now available to farmers on a commercial basis.

* The term 'proembryo' (*Henderson's Dictionary of Biological Terms*, 1979) is used to identify the conceptus from the time of fertilization until the appearance of the primitive streak, the cells of which give rise to the definitive 'embryo'.

As the first successful proembryo transfer was reported by Walter Heape as long ago as 1891, it is not surprising that many people have wondered whether the technique could be used in our own species, as a treatment for certain types of infertility. Yet it was not until 1979 that the painstaking work of Bob Edwards and Patrick Steptoe in England finally led to the birth of a baby conceived by *in vitro* fertilization (IVF) followed by transfer of the proembryo into its own mother's uterus. This technique (see Chapter 5) has now been taken up by centres all over the world, leading to the birth of hundreds of babies conceived by IVF.

Human proembryos, like those of other animals, can be successfully stored in liquid nitrogen. The first birth of a baby after freezing of the proembryo before implantation was reported in Australia in 1984. If several eggs are recovered from a woman's ovaries, it may be advantageous to store some in this way, so as to be able to transfer them in subsequent cycles. Storage also makes it easier to use donor eggs from another woman for transfer. Provided the eggs are recovered without any appreciable risk to the donor, there would seem no more ethical objections to this procedure than to donor insemination. Unfertilized eggs are taken from the donor's ovary during a sterilization operation and fertilized with spermatozoa from the patient's husband; alternatively, a proembryo that had reached the donor's uterus 3–4 days after fertilization (Table 6.1) could be flushed out through the cervix and transferred to the patient's uterus – a procedure referred to as 'proembryo transfer (donor)'. In either case the recipient, though not the genetic mother of the baby, would have

Fig. 6.1. Liquid nitrogen tank, with lamb born from a proembryo frozen in liquid nitrogen and subsequently transferred to the uterus of a foster-mother by S. M. Willadsen, ARC Institute of Animal Physiology, Cambridge. (From *The Freezing of Mammalian Embryos*, Ciba Foundation Symposium 52 (new series). Ed. K. Elliott and J. Whelan. Elsevier, Excerpta Medica; Amsterdam (1977).)

the satisfaction of carrying it through pregnancy. The contribution that each parent makes to a baby conceived in any of these ways is shown in Table 6.2. One circumstance in which egg or proembryo donation might be valuable would be for women with some genetic or chromosomal defect that they did not wish to pass on to their children, just as donor insemination may be used today where the husband has an hereditary disorder such as cystic fibrosis or Friedrich's ataxia.

Table 6.2 also includes an entry 'uterine fostering' to cover the case where husband and wife are both genetically normal and produce normal eggs and sperm, but where the wife cannot maintain a pregnancy. In

Table 6.1. *Eggs recovered from human oviduct and uterus surgically, at laparotomy, or from uterus by transcervical flushing*

		Eggs recovered from	
Days after ovulation	Oviduct	Uterus (surgery)	Uterus (flushing)
0	4/8	0/7	0/1
1	2/3	0/3	0/3
2	3/7	0/7	2/4
3	3/7	0/5	3/10
4	0/4	1/1	2/15
5	0/4	0/1	1/7
6	0/3	—	0/3

(Data from H. Croxatto. In *Physiology and Genetics of Reproduction*, Part B, p. 159 Ed. E. M. Coutinho and F. Fuchs. Plenum Press; New York (1974).)

Table 6.2. *Routes of parental contribution to children originating in various ways*

Origin of children	Paternal genes	Maternal genes	Uterine environ- ment	Postnatal care (maternal and paternal)
Normal	+	+	+	+
DI	−	+	+	+
IVF	+	+	+	+
PTD	+	−	+	+
IVF/DI	−	+	+	+
PTD/DI	−	−	+	+
Uterine fostering (surrogacy)	+	+	−	+
Adoption	−	−	−	+

DI = donor insemination; PTD = proembryo transfer (donor);
IVF = *in vitro* fertilization.

principle, the woman's egg could be fertilized *in vitro* with her husband's spermatozoa, and the resulting proembryo transferred to the uterus of the foster mother ('surrogate' mother). Such a procedure is not socially acceptable at present; ethical objections are that pregnancy is always accompanied by some risk to the health and even to the life of the pregnant woman, and secondly that the emotional bond formed between mother and baby before birth might prove too strong to allow the baby to be handed over to its 'genetic' parents after the birth without causing anguish to the foster-mother. The situation is not a new one, since throughout history men with infertile wives have sired babies by other women and then sought to adopt them. Occasionally, this arrangement works satisfactorily within a family, e.g. between two sisters, but attempts to offer financial inducements to the 'other woman' have sometimes ended in lawsuits for breach of contract.

Is it possible that the availability of proembryo transfer will in the future allow women to be paid to act as surrogate mothers merely to save others the bother of pregnancy? If that is what women choose, and if that is the way society is organized, then it is possible. In practice, most women I have talked to (especially those that have already experienced pregnancy and childbirth) value these experiences and would not lightly give them up. The social issue, since it is of wider relevance, will be discussed in the last section of this chapter.

Congenital defects

Many couples today choose to terminate a pregnancy rather than bring a malformed child into the world. If there is a higher than average risk of malformation, whether because one of the parents or an older child is affected, or just because the mother is older than usual, they may be offered antenatal diagnosis, in the form of amniocentesis, with the possibility of an induced abortion if the fetus proves to be abnormal. The procedure, which cannot be done until about 16 weeks' gestation, consists of removing a sample of amniotic fluid containing some free fetal fibroblasts. Little risk to the mother is involved, or to the fetus. A high level of alphafetoprotein (AFP) in the fluid indicates a severe neural tube defect (anencephaly, spina bifida) in the fetus. The level of AFP in the mother's blood tends also to be raised, so maternal blood sampling may be used as a preliminary screening procedure to identify women who should be offered amnio-centesis.

The fetal cells from the amniotic fluid can be cultured and chromo-some preparations made, to give information on the chromosome constitution of the fetus. Down's syndrome (mongolism, trisomy 21), one of the most common defects identified by amniocentesis, is recognizable by the presence of an extra chromosome, and, like some other autosomal trisomies (Table 6.3), shows a marked increase in incidence with maternal age (see Book 2, Chapter 5, and Book 4, Chapter 7). Chromosome sexing is sometimes used to identify male fetuses, which may be affected by a sex-linked genetic disease such as haemophilia. About 60 of the more than 1000 diseases known to be due to single gene defects can be diagnosed by analysis of enzyme activity in the fetal fibroblasts, but most of these conditions are very rare.

The much more common haemoglobinopathies (sickle cell anaemia and thalassaemia) cannot be diagnosed in this way, as fibroblasts do not synthesize haemoglobin. Fetal blood can be obtained from a placental vessel, and the red blood cells examined morphologically (a distortion of shape indicates sickle cell anaemia) or biochemically (low or absent synthesis of β-globin chains *in vitro* indicates β-thalassaemia – Fig. 6.2), but the procedure is much more difficult than amniocentesis and carries a 7 per cent risk to the pregnancy.

Table 6.3. *Maternal age is greater than average for trisomies, but not for other chromosomal abnormalities*

Karyotype	Maternal age (years) (mean $\pm$ SE*)
Normal	27.5 ± 0.4
Triploid	27.4 ± 0.8
Tetraploid	26.8 ± 1.4
Translocation	27.0 ± 2.3
XO	27.6 ± 0.9
Trisomy (all)	31.3 ± 0.6
Group A	29.6 ± 2.2
Group B	33.4 ± 7.1
Group C	30.9 ± 1.7
Group D	32.5 ± 1.3
Group E	29.6 ± 0.9
Group F	30.1 ± 5.3
Group G	33.2 ± 1.4

* SE = standard error.
(Data from J. Boué, A. Boué and P. Lazar. *Teratology*, **12**, 11–26 (1975).)

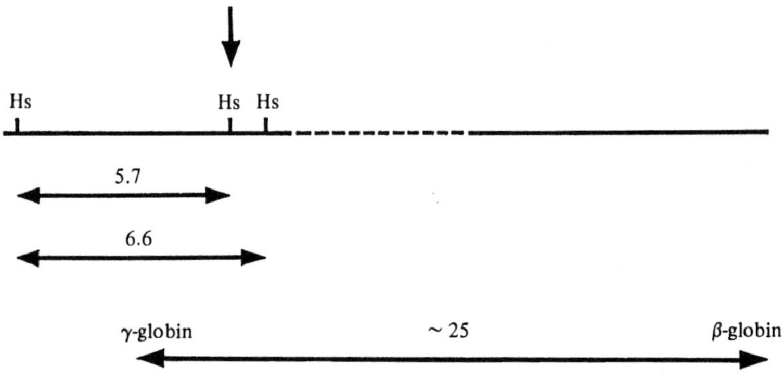

Fig. 6.2. Simplified diagram of structure of human globin gene complex. Three restriction sites for the restriction enzyme HsuI (Hind III) are marked; the arrowed site is the polymorphic one, present on some chromosomes but not all. Distances are in kilobase pairs.

An alternative approach to the diagnosis of genetic defects involves extracting the DNA from the fetal fibroblasts obtained by amniocentesis, and analysing it directly by the use of DNA hybridization and recombinant DNA techniques. A radioactive DNA probe, highly specific for the affected locus, may be used either to detect the gene defect itself, for example the gene deletions responsible for α-thalassaemia, or to show up a variation in a neighbouring restriction enzyme site that happens to be present on the same chromosome as the defective gene throughout most of a particular population. This strategy has been successfully used for the antenatal diagnosis of sickle cell anaemia. Alternatively, if a highly specific DNA probe is available for an adjacent gene (e.g. the gene for γ-globin, closely linked to that for β-globin), it may be used in the same way. The variant restriction enzyme site need not be confined to the same chromosome as the defective gene throughout the population, but can be analysed on a family basis, as was recently done by Bob Williamson and his colleagues in London for the antenatal diagnosis of β-thalassaemia (see Fig. 6.3). Since any coding gene, defective or normal, is likely to be surrounded by a unique set of DNA polymorphisms, this approach may prove widely applicable to the antenatal diagnosis of single gene defects.

Recently, prenatal diagnosis has been made earlier in pregnancy, with chorionic villi removed from the fetal part of the placenta between 8 and 12 weeks of gestation. The procedure is usually carried out through the cervix of the uterus, with ultrasound for guidance. Enough villi can be recovered for chromosomal or biochemical tests, or for DNA analysis. If the fetus is found to be abnormal, the pregnancy can then be terminated before the end of the first trimester, a much less difficult and distressing procedure than second trimester termination.

Paradoxically, the availability of antenatal diagnosis reduces the number of abortions requested, since couples with a family history of genetic

Fig. 6.3. Antenatal diagnosis for β-thalassaemia in an Indian family. DNA from the father, the mother, and a son homozygous for β-thalassaemia, together with DNA from amniotic cells of the fetus, was analysed for HsuI restriction enzyme sites in the γ-globin region. Autoradiography with a DNA probe showed that both parents and also the thalassaemic son had one chromosome with and one without the polymorphic site (DNA bands at 5.7 and 6.6 kilobase pairs respectively). Thus in one parent the β-thalassaemia gene must have been on the same chromosome as the polymorphic site. Since both chromosomes of the fetus lacked the site, it must have inherited a normal chromosome from one parent, so must be a carrier rather than a homozygous thalassaemic. (From P. F. R. Little *et al.*, *Nature*, **285**, 144–7 (1980).)

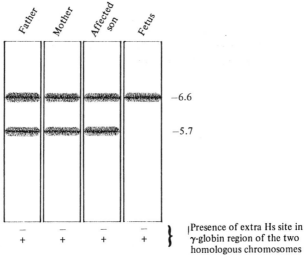

disease, or with one child already affected, often decide to terminate any future pregnancy unless they can be assured that the fetus is unaffected. Induced abortion, however, is a traumatic experience even early in pregnancy, and for many people it is unacceptable whatever the circumstances. Obviously, it would be preferable if congenital malformations, due to genetic defect or any other cause, could be prevented or cured rather than merely diagnosed before birth.

The immediate prospects are not very encouraging. Birth defects due to prematurity are becoming less common as we learn more about the proper care of premature babies. Some evidence suggests that the incidence of neural tube defects may be reduced by vitamin supplements. Infants born with an inherited inability to form antibodies, or with aplastic anaemia, diseases that are normally fatal, have recently been successfully treated with fetal liver or thymus or bone marrow cells. Donor and recipient must be carefully matched for histocompatibility type, so that the transplanted cells can become established and set up a state of permanent chimaerism compensating for the deficiency in the recipient. Perhaps other forms of anaemia will eventually be treated in a similar way. We know from studies on mice that blood chimaeras of this sort can be made at any age, but that if chimaerism is to be achieved in tissues other than blood, the donor cells must be introduced much earlier in development. Cells carrying gentic defects that would be lethal before birth on their own can survive throughout life if introduced at the blastocyst stage.

From a long-term point of view, however, it is undesirable merely to provide a cellular crutch that will enable genetically defective embryos to survive and possibly reproduce. Recombinant DNA techniques coupled with methods of gene transfer may eventually enable us to replace defective genes with their normal counterparts. If this could be done in the fertilized egg, possibly with the aid of nuclear transfer (see next section), genetically defective individuals would be able to reproduce without fear of transmitting their abnormality to future generations. Genetic defects can be cured only by genetic engineering.

Twenty-first century reproduction?

Choosing the sex of your baby

An effective female-determining or male-determining method that would ensure that progeny were of the desired sex has been a dream of farmers for many years. The dairy cattle industry wants mainly heifer calves, while beef cattle breeders prefer an excess of bull calves; either way, the potential economic rewards are enormous.

In our own species, the possibility has not attracted much public attention until recent years, when in many parts of the world family size has been reduced to two or three children. In China, couples are now under pressure to have one child only. The fewer children a couple have, the more important it is to them that they should be healthy (see last section) and of the desired sex. Most people questioned in Britain or the USA would like two children, one of each sex. Yet by chance, 50 per cent of all two-children families will consist of one sex only. Every few years the popular press announces that an infallible way of choosing the sex of your child has at last been discovered. What is the present situation?

Many societies have controlled the sex ratio of their children by infanticide, killing or leaving to die a proportion of babies of the 'wrong' sex (as far as I know, always girls). This solution is effective, but not socially acceptable today. Amniocentesis, or even direct visual examination by fetoscopy, allows the sex of a baby to be known midway through pregnancy, at a stage when induced abortion is still possible, while recovery of chorionic villi (see page 183) makes it possible to determine the sex of the fetus earlier, during the first trimester. Families afflicted by one of the rare genetic diseases that affect one sex only may choose to make use of one of these methods, allowing the pregnancy to continue to term only if the fetus is of the unaffected sex. In principle, amniocentesis could be used for selecting on non-medical grounds, and has sometimes been so used, for example in India, but few women having got midway through a pregnancy would wish to terminate it merely because the baby was not of the desired sex, and few doctors would be prepared to perform an abortion for such a reason. Both these arguments would weaken if the chorionic villus method proves safe and reliable, since during the first trimester many terminations are carried out for non-medical reasons anyway, legally or illegally, all over the world. However, we do not yet know how great is the risk of causing abortion by mistake with this procedure.

Proembryos of farm animals grow to a much larger size before implantation than do those of humans or mice, which means that a piece of trophoblast can be removed from the blastocyst for chromosome sexing without affecting development. So, at the cost of a little extra work, proembryo transfer in cattle could produce male or female calves to order. Some analogous technique could perhaps be devised for human preimplantation proembryos, by removing one or two cells for chromosomal or immunological analysis during cleavage. For the small minority of women conceiving by IVF, it would then become possible to replace in the uterus only proembryos of the desired sex.

What about artificial insemination? Since the sex of a proembryo is decided by whether the spermatozoon fertilizing the egg carries an X- or a Y-chromosome, separation of X-bearing from Y-bearing spermatozoa should enable each sperm population to be inseminated separately, according to whether female or male offspring were desired. In cattle, where artificial insemination is so widely used, such a facility would be of great economic value, and in our own species it would certainly be appreciated by patients receiving donor insemination; but despite claims to the contrary, it has still not been achieved.

Advances in any of these areas could be quite rapid, leading to clinical applications within the next decade. What seems unlikely to emerge before the twenty-first century, since no leads exist at present, is a universally applicable method of regulating sex at conception, perhaps by using some substance that could be taken orally by either partner before coitus, or applied to the vagina. Apart from widening the area of choice, the impact of such a method would not be so great in Britain or the USA, since most couples would like one child of each sex; but in countries where male children are valued more highly than female, the effect on the sex ratio might be considerable. If the method were freely available, a large excess of boys would be produced initially, which would presumably increase the value set on girls. If it were available only to the rich, a distortion in marriage patterns would be predicted: more rich men than usual would marry into families poorer than their own.

Cloning

Another topic that is still in the realms of science fiction as far as our own species is concerned, but which could become science fact in the twenty-first century, is cloning. A clone is a group of genetically identical individuals; any pair of monozygotic twins is thus a clone of two. In mice, sheep, cattle and horses, monozygotic twin pairs have been made experimentally, by splitting the proembryo into two parts during early cleavage and allowing each to develop independently. In sheep, even monozygotic quintuplets have been made by this technique (see Fig. 5.14); the quarter or half proembryos grow to full normal size, and offer a useful experimental means of analysing the relative importance of nature and nurture (see Book 2, Chapter 1). In our own species, monozygotic twinning occurs spontaneously at a relatively high rate; there seems no reason to increase its rate experimentally.

An alternative method of producing clones of animals is by nuclear transfer. In amphibia, the nucleus of a fertilized egg can be replaced by a donor nucleus taken from a proembryo or from an adult animal. When an embryonic nucleus is used, the egg will develop successfully right through to the adult stage (see Fig. 6.4), and since many nuclei can be taken from the same donor proembryo, a clone of several dozen genetically identical frogs can be produced. When the donor nucleus is taken from an adult animal, development is arrested at or before the tadpole stage; we do not yet know whether the failure to produce frogs reflects some true restriction of developmental potency on the part of the nucleus, or whether it is due merely to some as yet unidentified technical problem. In mouse eggs undergoing fertilization, pronuclei can be removed and replaced by pronuclei from eggs of a different strain without affecting development. There is also a report that nuclei from the pluripotent inner cell mass cells will support development when substituted for pronuclei in eggs undergoing fertilization, but this work has not yet been repeated. As far as adult donors go, no nuclear transplantation of adult nuclei to pronucleate eggs has been

Fig. 6.4. Transplant procedure starts with preparation of a frog's unfertilized egg for receipt of a cell nucleus by destroying its own nucleus through exposure to ultraviolet radiation. Next, intestine is taken from a tadpole that has begun to feed and cells are taken from its epithelial layer. A single epithelial cell is then drawn into a micropipette; the cell walls break, leaving the nucleus free. The intestine-cell nucleus is transplanted into the prepared egg, which is allowed to develop. In some 1 per cent of transplants the egg develops into a frog that has one nucleolus in its nucleus instead of the usual two. (From J. B. Gurdon, *Sci. Amer.* **219**, 24–35 (1968).)

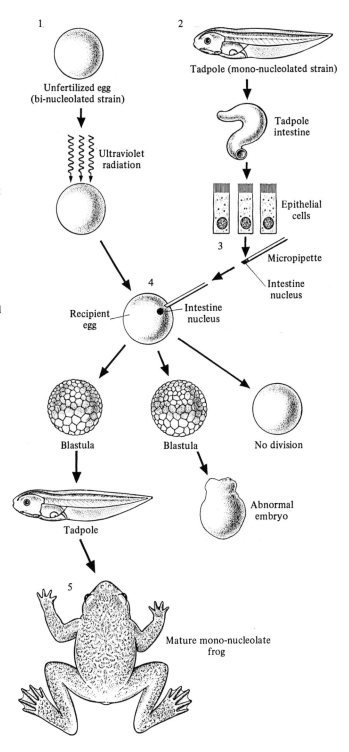

successful, and indeed there is an increasing amount of evidence to challenge the dogma that the genetic constitution is the same in all cells of the body. The differentiation of mammalian lymphocytes (see Book 4, beginning of Chapter 6, Second Edition) affords one example of gene rearrangements during development.

Thus, although nuclear transfer has now been shown to be possible in mammals as well as in amphibia, what evidence there is argues against the feasibility of the 'science fiction' concept of cloning, in which nuclei are taken from the body cells of an adult individual (typically an elderly male millionaire!) in order to produce one or many children 'in his image'. With or without this restriction, what possible applications for cloning could or should be envisaged if the appropriate techniques were extended to our own species in the twenty-first century, and what ethical objections might be envisaged?

There seems no widespread objection to having small numbers of 'identical' human beings around; there is no move to suppress one of each pair of monozygotic twins at birth. Indeed, anyone who has known monozygotic twins realizes how very far from 'identical' so-called genetically identical people in fact are. It may be that the objection is not to identicalness, but to the idea of 'copying' an existing human being by nuclear transplantation, if this could ever be done. Would it be more or less ethically acceptable for a sterile couple to have two children by donor insemination, donated proembryo, or a combination of the two if both parents were infertile, with the possible ethical and psychological problems of involving a 'third party' in the marriage, or to have two 'nuclear transplant' children, a boy from the father and a girl from the mother? Proembryo aggregation might even be used to mingle the contributions of the two parents and avoid too close a similarity to either.

An alternative type of ethical objection involves the size of the clone. Maybe two or three very similar human beings are all right, but not hundreds or even dozens. In fact, even if the idea of such a clone of people came to be acceptable, it is hard to see what purpose it could serve in our present social context. Jobs may exist that would be better done co-operatively by genetically identical than by unrelated people, but the difference cannot be great or monozygotic twins would probably already be employed in this way. Similar upbringing seems more likely than similar heredity to produce people who can work well together.

One potential application of nuclear transfer methods in our own species lies in the cure of genetic defects, mentioned briefly earlier in this chapter. For example, if cells from genetically defective proembryos were cultured when the pregnancy was terminated, recombinant DNA and gene transfer techniques might be used to correct the faulty genes and an improved nucleus could then be transferred to another fertilized egg from the same couple. Whether it would ever be economically justifiable to devote so many resources to curing a single defective proembryo is a separate question.

Parthenogenesis

The eggs of mammals can be induced to develop in the absence of a fertilizing spermatozoon, by various chemical and physical stimuli including calcium-free medium, alcohol, heat and electric shocks. Such parthenogenetic embryos are, of course, all female; they may continue development apparently normally until mid-gestation, but they have never as yet survived until birth, even when the single haploid chromosome set of the egg has been doubled up to give the normal diploid number of chromosomes. We do not yet know why they die.

Because birds have a different chromosomal sex-determining mechanism from mammals, parthenogenetic bird embryos are all males. Strains of turkeys and chickens are known in which occasional eggs undergo spontaneous parthenogenetic activation, some even developing into adult birds (Fig. 6.5).

Could human eggs be induced to develop parthenogenetically? In an outbreeding, genetically heterogeneous species such as our own, any technique based on doubling up a single set of haploid chromosomes would be rapidly lethal owing to the uncovering of harmful recessive genes, however well it might work in inbred mice. Parthenogenetic human embryos would therefore die even earlier than similar mouse embryos, unless the first meiotic division leading to the formation of the egg could be suppressed. Instead of a single haploid set, the egg would then contain the full complement of maternal chromosomes, as do other cells in the mother's body.

Fig. 6.5. Mature parthenogenetic male turkey reared by Dr. M. W. Olsen, US Department of Agriculture, Beltsville, Maryland. This bird was fertile, with equal proportions of males and females in his progeny. (From R. A. Beatty. Parthenogenesis in vertebrates. In *Fertilization*. Ed. C. B. Metz and A. Monroy. Academic Press; New York (1967).)

Vitriparity

Aldous Huxley in *Brave New World* envisaged a future in which development from conception right through to 'birth' took place in tanks of culture medium, with the composition of the medium determining whether the baby would end up as an alpha or beta, or as a delta or epsilon destined to do only the most boring of tasks. Fortunately (in my view), research has gone in a direction that Huxley never predicted, so that the problem of boring repetitive jobs is more likely to be solved by making machines intelligent than by making people stupid.

Will it ever be possible to achieve embryonic development entirely *in vitro*, so that we shift from viviparity to vitriparity? Mouse proembryos can now be cultured from fertilization up to a fairly normal early somite, beating-heart stage, which represents nearly half of the entire gestation period in a mouse, but in a human would account for less than 10 per cent of development. After that stage the mouse proembryo gets too big to manage without a placenta, and no one has yet devised a satisfactory artificial placenta. As far as we know, there is no basic biological reason why development could not take place entirely outside the uterus: the problem is a technological one, largely a question of plumbing, and could no doubt be solved fairly rapidly, at least for the mouse, if a sufficiently large amount of money and effort were to be deployed.

The knowledge gained in the course of the work would certainly advance our understanding of the causes of birth defects, and the interactions between mother and implanted embryo in normal pregnancy. But to adapt such technology from the mouse to a species such as our own, with a large fetus and a protracted gestation period, would be a colossal task. At present there seems no reason to think it would ever be attempted; even if vast resources were to be devoted to it, the chances of achieving a successful result before the end of the twenty-first century would be slim.

Choice or chance?

Most people in Britain and America today have far more choice, far more say in their lives than they would have had 100 years ago, let alone 500 years ago. Food, drink, clothing, jobs, education, housing, holidays, marriage partner – in spite of economic recessions, unemployment and inflation, the inhabitants of developed countries still have a dazzling range of alternatives to choose from. From an evolutionary point of view, reproduction is always the most conservative activity of a species, more so, for instance, than food preferences or habitat selection; so perhaps it is not surprising that in our own species, too, reproductive activity is one of the last areas that cultural evolution has brought under our conscious control.

Some of the possibilities that I have outlined in this chapter may strike the reader not only as bizarre, but also as repugnant, and from the ethical point of view totally unacceptable. We must, of course, be free to reject them; but we must also remember that our grandchildren and great-grandchildren may not feel as we do. Few of us would wish to be bound by the ethics of the Middle Ages, when good and worthy people subjected witches to ordeal by water and burned heretics at the stake. Few would feel wholly at ease with nineteenth century ethics. Why then should we imagine that our twenty-first century descendants will view life in the same ethical light as we do?

Are we on the other hand in danger of condemning our descendants to a future in which all reproduction is by donor insemination from officially selected sperm banks, or in which 90 per cent of women are allowed no children of their own but bear those of the remaining 10 per cent, or the cloned progeny of a mere 0.1 per cent? The way to avoid such chilling prospects is to reject the type of social system that would enforce them, not to reject the biomedical knowledge that it misuses. It is a commonplace that all scientific advances can be used for evil as well as for good. The same microbiological discoveries that prevent epidemics provide a possible basis for bacteriological warfare; the discovery of fire, crucial for human civilization, made possible the Dresden fire-storm; surgery can be used to cure or to deform. As Epictetus wrote nearly 2000 years ago: 'Everything has two handles, by one of which it ought to be carried and by the other not.'

Medical research today cannot be carried out without teams of scientific workers, and large sums of money. Most of this money comes from government agencies, so in a democratic society the research that is done, as well as the way in which scientific discoveries are used, can be influenced by public opinion. Public concern about the way in which heart trans-plantation was being conducted in Britain resulted in the work being halted for several years, and led directly to the introduction in 1976 of the 'brain death' code of practice. In recombinant DNA research, a moratorium was declared and all experiments stopped for a period, until the risk that new infectious agents might accidentally be created could be assessed. The risk is now known to be small, and the potential economic benefits very large, so the work is going ahead again but with strict safeguards.

To hold up research in any field for too long may have its own hazards. The value of a piece of knowledge, like that of a piece of string, depends on circumstances, but may turn out to be very great. Even if our species never decides to embark on vitriparity, knowledge of how to grow the human conceptus safely outside the uterus would be worth having. Indeed, in the unlikely event of some hitherto unknown 'contagious abortion' epidemic sweeping the world, such knowledge might be crucial to the survival of our species.

Doctors and scientists have a responsibility to make as widely known as possible the advances in knowledge that research is making, and the implications that these have for possible innovations in medical technology.

But decisions as to *how* the new knowledge should or should not be used, and how much of society's resources should be spent in any particular area, such as alleviating infertility, are essentially political, and in a democratic society should involve the entire community.

In a future with fewer jobs and shorter hours of work, the one commodity in plentiful supply will be leisure. If some of this leisure is used for education, people will be able to get the knowledge they need to make choices and take decisions on matters affecting their own lives and the future of their families. They may go further, and bring democracy to life by actively participating in decisions affecting the future of society. They may even acquire the power to choose how much choice they should have, and thus inaugurate what I believe to be the next stage of human cultural evolution.

Suggested further reading

In vitro fertilization and embryo transfer in human beings. J. D. Biggers. *New England Journal of Medicine*, **304**, 335–42 (1981).

Genetic Manipulation: Impact on Man and Society. Ed. W. Arber, K. Illmensee, W. J. Peacock and P. Starlinger. Cambridge University Press (1984).

Artificial Insemination. Proceedings of the Fourth Study Group of the Royal College of Obstetricians and Gynaecologists. Ed. M. Brudenell, A. McLaren, R. Short and M. Symonds. RCOG; London (1976).

The Family and its Future. Ciba Foundation Symposium. Ed. K. Elliott. Churchill Livingstone; London (1970).

The Freezing of Mammalian Embryos. Ciba Foundation Symposium 52. Ed. K. Elliott and J. Whelan. Elsevier, Excerpta Medica; Amsterdam (1977).

Man and his Future. Ciba Foundation Symposium. Ed. G. E. W. Wolstenholme. Churchill Livingstone; London (1963).

Law and Ethics of AID and Embryo Transfer. Ciba Foundation Symposium 17. Ed. G. E. W. Wolstenholme and D. W. Fitzsimons. Elsevier; Amsterdam (1973).

A Matter of Life. R. G. Edwards and P. C. Steptoe. Hutchinson; London (1980).

Report of the DHSS Committee of Inquiry into Human Fertilisation and Embryology. Chairman Dame Mary Warnock. Her Majesty's Stationery Office; London (July 1984). Now published as a book: *A Question of Life.* Blackwell Scientific Publications; Oxford (1985).

The New Genetics and Clinical Practice. D. J. Wetherall. The Nuffield Provincial Hospitals Trust (1982).

Human Embryo Research: Yes or No? Ciba Foundation Study Group. Tavistock Press; London (1986).

7

Barriers to population control

C. R. AUSTIN

Be it from idle curiosity or genuine concern, anyone interested in reproductive biology must surely give some thought to the consequences of unabated human fecundity. Indeed, the matter has exercised the minds of many people for many years. Traditionally, we attribute the first clearly stated forebodings to Thomas Malthus who wrote, in 1798: 'Population, when unchecked, increases in a geometric ratio. Subsistence increases only in an arithmetic ratio.' And on this premise he saw catastrophe ahead. Quite immediately ahead, too, for he predicted that, unless drastic controls were exercised, the population of Britain was likely to grow from the 7 million of his day to 112 million by 1900, while the capacity for food production would by then have increased only enough to support 35 million people. In these circumstances, the dire alternative to mass starvation seemed to be mass emigration, but really there was no 'solution' since the rest of the world's population would be outstripping its means of subsistence in the same manner.

Unperceived by Malthus, there was for Britain and other countries of the Western World a third possibility, a kind of population maturation: environmental factors were to bring about a demographic transition in the nineteenth century. It was not the first occasion in human history for a phenomenon like that to happen, but the timing in this case was singularly fortunate. As a result, population growth in these countries slowed to a virtual standstill, or even a slight decline in some instances, and public attention shifted to other worries, such as nuclear warfare, deteriorating international relations, deficiencies in world health, and the difficulties of conserving environmental quality, limited natural resources and endangered species. Basically, however, the problem had not changed, for the underlying cause for those anxieties was and continues to be the massive population growth in developing countries, where numbers that would have been considered astronomical to the point of fantasy early in the century were fast becoming reality. Despite strenuous corrective efforts by national and international bodies, this unruly giant continues to grow, apparently immune to the devices of modern science.

There seem to be inherent barriers to population control in the developing world, its power of multiplication attesting to the process of reproduction as a basic function of life, a driving force of great strength, sufficient to overcome almost any opposition and ensure the survival of the species. Could not this be the reason why humanity's efforts at artificial control have so far proved futile? But why does the developing world not experience the same kind of maturation that saved the West from becoming too big for its boots? And what will happen if population growth inexorably proceeds as it is going now? In this chapter I will try to work out answers to these questions.

Population growth – past, present and future

The world scene

Population growth from the time of the earliest records to the present day has taken the form of an exponential curve (Fig. 7.1), indicating that the rate of increase has accelerated. The process is also clearly shown in Table 7.1, where the time for each increment of 1000 million gets progressively shorter. Neither the illustration nor the table offers strong support for the optimistic view that the *rate* of increase for the world is slowing down, but such generalizations are difficult to make anyway. Regions and countries differ from one another. Thus, over the past 30–35 years the world trend has actually been slightly downwards (Table 7.2), but whether this is a significant shift is questionable. On the other hand, there can be little doubt that a real reduction has taken place in the developed countries and East Asia, the change for the latter being attributable largely to the striking success of the Chinese efforts since 1970. The other developing countries have shown no great change, except for Africa which has actually increased by nearly 50 per cent. When we look at crude birth rates, the regional and country differences are more distinct (Table 7.3). Here there is no doubt that the rates for most countries and regions show a definite drop (the East Asian figures again due mainly to the Chinese results). Africa stands out

Fig. 7.1. How the world population has increased until the present day, and how it could grow in the next 20 years or so. (From World Census figures and the calculations in Table 7.1.)

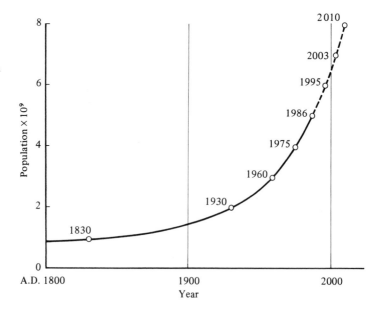

Table 7.1. *World population projections depend essentially on when the rate of increase ceases to accelerate, as it is doing at present, and begins to decelerate. The change would mark the point of inflection of the population curve and this could occur about 1986, with a total population of 5×10^9 and a rate of increase such that 1000 million people are added in 11 years (columns 2 and 3). If the population curve is symmetrical about its point of inflection, numbers could stabilize soon after the year 2131, with a total of a bit more than 8×10^9. But if the point of inflection occurs later, in A.D. 1995, 2003 or 2010, when only 9, 8 or even 7 years suffice for the addition of 1000 million people, then the corresponding numbers at 'stability' could well be something in excess of 10, 12 or 14×10^9 people*

Total population	Years taken	A.D.	Years taken	A.D.	Years taken	A.D.	Years taken	A.D.
1×10^9	say, 2×10^6	1830						
2	100	1930						
3	30	1960						
4	15	1975						
5	**11**	**1986**	11	1986	11	1986	11	1986
6	15	2001	**9**	**1995**	9	1995	9	1995
7	30	2031	11	2006	**8**	**2003**	8	2003
8	100	2131	15	2021	9	2012	**7**	**2010**
9			30	2051	11	2023	8	2018
10			100	2151	15	2038	9	2027
11					30	2068	11	2038
12					100	2168	15	2053
13							30	2083
14							100	2183

as the exception, with no real change in birth rates. All these figures tend to support the contention that the growth curve for the world as a whole has reached, or is approaching, the point of inflection, and that we might look forward to a diminishing rate of population increase from now on.

But predictions about population trends are notorious for their inaccuracy. In the 1930s and early 1940s, several well-known demographers made forecasts on the future course of the US population curve (Fig. 7.2) and proved astonishingly wide of the mark. The difficulty they encountered remains to this day. In an address given in July 1984, A. W. Clausen, President of the World Bank reported that the 'standard' projections of the Bank indicated that the world population would stabilize at a little more than 11000 million by the year 2150, but admitted that the assumptions made may well have been over-optimistic. Clearly, several

Table 7.2. *Annual rates of population growth (per cent)*

Region	1950–55	1960–65	1970–75	1975–80	1980–85
Africa	2.11	2.44	2.74	3.00	3.01
Latin America	2.72	2.80	2.51	2.37	2.30
East Asia*	2.08	1.81	2.36	1.47	1.20
South Asia	2.00	2.51	2.44	2.30	2.20
Developing countries	2.11	2.30	2.46	2.14	2.02
Developed countries	1.28	1.19	0.89	0.74	0.64
Total world	1.84	1.96	2.03	1.77	1.67

* Excluding Japan.
(From R. S. McNamara (1984), see Suggested further reading – UN 1982 assessment.)

Table 7.3. *Crude birth rates (per 1000 population)*

Region	1950–55	1960–65	1970–75	1975–80	1980–85
Africa	48.1	48.3	47.0	46.9	46.4
Latin America	42.5	41.0	35.4	33.3	31.8
East Asia*	45.0	37.3	33.9	22.3	18.8
South Asia	45.6	45.8	40.6	37.7	34.9
Developing countries	45.4	42.8	38.7	33.5	31.2
Developed countries	22.7	20.3	17.0	15.9	15.5
Total world	38.0	35.9	32.7	28.7	27.3

* Excluding Japan.
(From R. S. McNamara (1984), see Suggested further reading – UN 1982 assessment.)

possibilities should be kept in mind, and a useful way of doing this is to consider the consequences of taking four different dates for the point of inflection to occur, with curves symmetrical about those figures. On this basis, and with the point of inflection located near the present (the year 1986 is selected), the population should reach equilibrium at about 8000 million in about the year 2130 (Table 7.1). Other possibilities are also dealt with in the table. In each instance, equilibrium is taken to be reached when the time required for the next increment of 1000 million exceeds 100 years. If the point of inflection does not materialize until early in the next century, the world seems likely to be graced with 12000–14000 million people in the latter part of the twenty-second century.

At present, we are faced with the virtual inevitability of a world population about 20 per cent higher than it is today by the year 2000. This means that living space, food, health services, education, employment, resource use and pollution control will have to meet the needs of an additional thousand million people (the present population of China) in the next 14 years. Given a large measure of international goodwill, and the fact that technological progress also proceeds exponentially, it should

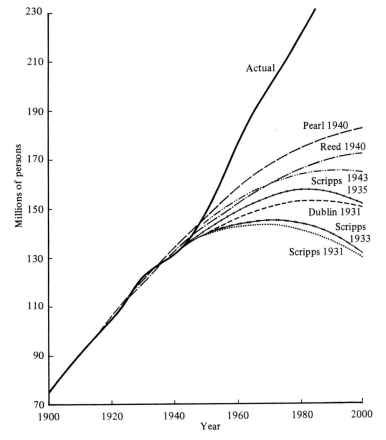

Fig. 7.2. Predictions made by leading demographers in the 1930s and early 1940s concerning the future trend in the US population. (Fig. 11-11 in J. L. Simon (1981) – see Suggested further reading.)

not be impossible, but of course it does represent one of the greatest challenges yet faced by man.

Population maturation

From late in the eighteenth century to the early part of the twentieth, the birth rate of Western nations fell quite dramatically – by more than a half. This was, of course, well before the advent of modern contraceptive agents and methods, so reliance was placed chiefly on the condom and coitus interruptus, with a back-up of abortion. But many socially significant events occurred during that time, which led to great improvements in health care, including control of several specific causes of death, which in turn increased general life expectancy. There were also improvements in

Fig. 7.3. Advances in public health and social conditions generally during the past 200 years, mostly as they affected life in the UK. (Modified from *History of the Homeland; the Story of the British Background.* H. Hamilton. Allen & Unwin; London (1947).)

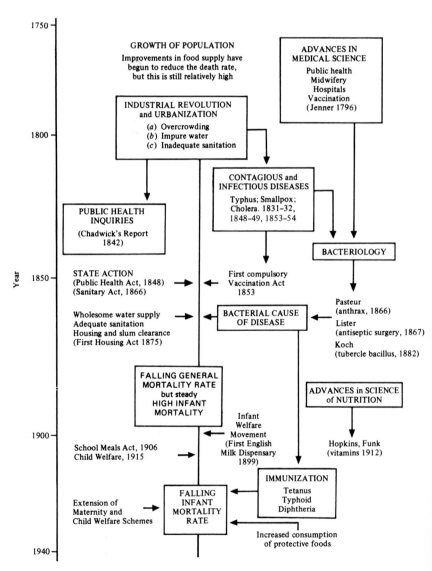

standards of living, education and social stability, equity and justice. Some of the highlights of this transition, as it affected life in Britain, are summarized in Fig. 7.3. The important consequence of biological significance was the reduction in the death rate, for after a latent period of about 50 years the birth rate declined steadily; this association of events is well shown by the vital statistics of Sweden (Fig. 7.4). The overall pattern of interrelated changes can be referred to as population maturation (Fig. 7.5), in which the causally connected trends of birth and death rates constitute

Fig. 7.4. Trends of birth and death rates in Sweden from 1720 to 1960; illustrating the demographic transition. (Fig. 11–8 in J. L. Simon (1981) – see Suggested further reading.)

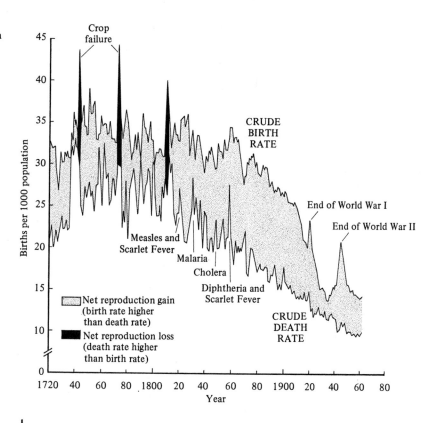

Fig. 7.5. Components of the demographic transition. A–B: birth and death rates both high, but the somewhat higher birth rate sustains a gradual population increase. B–D: death rates decline and population growth accelerates accordingly. C–D: birth rates now decline and so the population curve approaches a plateau.

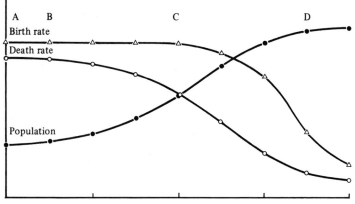

the demographic transition. But we should not underestimate the role of other components of the maturation picture, the upgrading of the social environment.

In several developing countries the death rate has now been significantly reduced, but the birth rate remains high, and at first one is tempted to wonder why the demographic transition does not ensue, as in the West's experience. In part it is because the time-scale is quite different: with the West, the death rate was brought down over a period of some 200 years, and the birth rate followed suit after an interval of about 50 years. In the developing world of today, the death rate has been reduced much more rapidly, mostly in the first half of this century, by the application of advanced Western technology in medicine and public health, and there has hardly been time enough for the birth rate to respond. Of course in some areas the death rate has not been brought down, either because of lack of medical attention or because of prolonged drought leading to mass starvation; there has been a great increase in death rate, in fact. But an additional important factor undoubtedly lies in the persistence of relatively primitive social conditions. Social instability, too, must exert its influence: in the post-colonial era, particularly in Africa, considerable unrest has been associated with rivalry between tribes; the colonial divisions of the country paid little heed to tribal associations, so there was a great deal of reshuffling to be done in the early years of independence.

The fall in the birth rate in the Western demographic transition has a special interest, for it could well represent the operation of an inherent mechanism for fertility control. We have already seen evidence of this process in the reproductive patterns of K-selected species of mammals (discussed in Book 4, Second Edition), relatively large animals that regulate their prolificacy with due regard to the limitations of their environment. It is not a perfect mechanism by any means, but K-selected animals do generally manage to avoid having the kind of 'boom-and-bust' population pattern so distinctive of rodents and other, generally small, r-selected beasts. And man is a member of the K group so should possess this kind of 'governor'. The demographic transition seems to provide evidence for its possession.

Rescue of the developing world
One of the best opportunities ever offered to countries of the developing world to discuss their problems and define solutions was the International Conference on Population, held in Mexico City in August 1984. Representatives from more than 140 countries and a number of funding agencies met and exchanged views, and agreed a Declaration of Mexico, with 88 recommendations for the continued implementation of the World Population Plan of Action drawn up at Bucharest in 1974. This was done with the recognition that 95 per cent of the projected world population increase to the year 2000 would occur in the developing world (Fig. 7.6), and there

was almost complete agreement that population control was essential if the developing countries were to achieve a balanced economy and a satisfactory standard of living within a reasonable time-span.

The Declaration urged recognition of the need for proper utilization of natural resources and protection of the physical environment, and of the need for action to attack the problems of disease, mass hunger, illiteracy and unemployment; to provide everyone with the information, education and means to be able to decide freely and without coercion the number and spacing of their children; to improve the status of women and the levels of maternal and child health; to develop better contraceptive methods without promoting abortion as a method of family planning; to expand research into problems of infertility and sub-fertility; to enable countries to achieve self-reliance in the management of their population programmes; and to integrate rural and urban development strategies in the light of the knowledge that nearly half of the world's population would be living in cities by the end of the century.

These recommendations will, of course, require the expenditure of vast sums of money, and the developing countries as a whole are notably short of financial resources. The funds must come from the developed world for that is where most of the wealth is. The requisite contribution will call for an enormous commitment on the part of the developed countries, well in

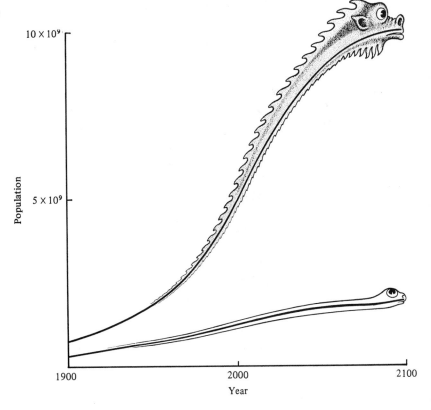

Fig. 7.6. From now on, the growth of population will be much greater in the developing world than in the developed world.

excess of anything that has been managed so far, for food, medical supplies, equipment and expertise, but particularly for money, because of its flexibility. The aid must be provided *unconditionally*. The importance of this point was made abundantly clear at the Conference, when the USA demanded that recipient states adopt a free market economy in the Western style. There was almost total rejection of this point of view; delegates from country after country, protesting at what they saw as an attempt to impose a particular economic philosophy, pointed out that poor countries could not hope to repeat the economic history of Western nations in the time available. As the representative from Kenya said, 'We cannot wait for economic growth to put everything right. We cannot wait 150 years.' It is also important that financial contributions should carry only a very low rate of interest or be actual gifts. Ingrained in Western thinking is the idea that loan making should be profitable; the cynicism underlying much of Western 'charity' is profound.

That we need a new philosophy in dealing with the world's economic problems was one of the recommendations of Willy Brandt and his Independent Commission, composed mostly of representatives from developing countries. They crystallized their conclusions with the aid of a map (Fig. 7.7) based on the Peret projection, which gives a truer indication of the actual sizes of countries than does Mercator's projection, and a meandering line separating 'North' from 'South'. Their report (see Suggested further reading) presents a thoughtfully argued and detailed case for the North to come to the aid of the South, a move they consider to be an overriding and urgent necessity.

Fig. 7.7. A map of the world based on Peret's projection, which depicts more accurately than Mercator's and other commonly used projections the relative areas of land masses. The shaded areas constitute the 'North' (the developed world) and the unshaded the 'South' (the developing world). (From *North–South: a Programme for Survival* – see Suggested further reading.)

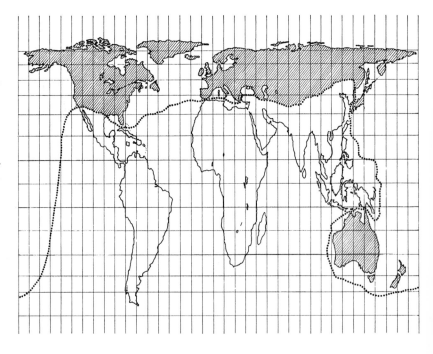

William Clark (see Suggested further reading) has resumed the attack, maintaining that the countries of the developed world had indeed begun to tackle the problem in something approaching a realistic way, but had become daunted by the magnitude of the task and so now strove with diminished motivation. Financial difficulties at home were a major factor; nevertheless, at a meeting of 21 donors of the International Development Association in 1984, 20 voted in favour of increasing aid to the poorer countries. The one that voted against was the USA, which is ironic, because the current American GNP *exceeds* that of Japan, West Germany, France, UK, Italy, Spain, Netherlands, Switzerland, Sweden and Portugal *put together*. A wry commentary on the situation took the form of the cartoon shown in Fig. 7.8. Such a policy prompted one of Clark's recommendations, namely that what is needed is some kind of world central bank that would have the power not just to ask for help but actually to *tax* the wealthier countries.

Some attention should be given at this point to population growth rates in specific countries. The basic information is set out in Table 7.4, and in Table 7.5 we see the population doubling times corresponding to the various growth rates. The highest rate was for the Ivory Coast, namely 5.0, which means a doubling time of 14 years. But that is a small country; more ominous are the figures for countries like Brazil and Mexico, which are due to double their already large populations in a little over 20 years. As we should note, however, the prize must go to India, due to become the most populous country within the next hundred years (Fig. 7.9). Western countries are almost static in terms of population growth, and the

Fig. 7.8. A comment on the world's economic dilemma. (From *Encyclopaedia Britannica Book of the Year 1984*, p. 315.)

Table 7.4. *Populations and growth rates for countries with more than three million inhabitants*

Country	Number (mid 1982) × 10³	Growth rate (av. 1970–81) (%)
China	1008175	1.5
India	698000	2.1
USSR	270000	0.9
USA	231990	1.0
Indonesia	151720	2.3
Brazil	124343	2.1
Japan	118460	1.1
Bangladesh	92619	2.6
Nigeria	89118	2.5
Pakistan	87125	3.0
Mexico	73011	3.1
West Germany	61713	0.0
Italy	56224	0.4
UK	55782	0.1
Vietnam	55503	2.8
France	54257	0.5
Philippines	50740	2.7
Thailand	48450	2.5
Turkey	46312	2.3
Egypt	45000	2.5
Iran	40476	3.1
South Korea	39331	1.7
Spain	37935	1.1
Burma	37065	2.2
Poland	36062	0.9
Ethiopia	32580	2.0
Zaire	30369	3.0
Argentina	28438	1.6
Columbia	27098	1.9
South Africa	25687	2.8
Canada	24625	1.2
Yugoslavia	22646	0.9
Romania	22638	0.9
Morocco	20420	3.1
Algeria	19954	3.3
Sudan	19435	3.1
Tanzania	19111	3.4
North Korea	18789	2.6
Taiwan	18735	1.6
Peru	17401	2.6
Kenya	17142	4.0
Afghanistan	16786	2.5
East Germany	16732	−0.2
Nepal	15769	2.6
Czechoslovakia	15375	0.7
Sri Lanka	15189	1.7

Table 7.4. (*cont.*)

Country	Number (mid 1982) × 10³	Growth rate (av. 1970–81) (%)
Australia	15054	1.4
Venezuela	14714	3.4
Malaysia	14344	2.5
Netherlands	14286	0.8
Iraq	14014	3.4
Uganda	13651	2.6
Mozambique	12615	4.2
Ghana	12244	3.0
Chile	11275	1.7
Hungary	10702	0.4
Portugal	10056	0.8
Belgium	9855	0.2
Cuba	9799	1.1
Greece	9793	0.9
Syria	9413	3.7
Madagascar	9400	2.6
Bulgaria	9108	0.5
Ivory coast	8938	5.0
Saudi Arabia	8905	4.5
Cameroon	8853	2.2
Sweden	8330	0.1
Ecuador	8073	3.4
Guatemala	7699	3.1
Austria	7571	0.1
Zimbabwe	7540	3.2
Mali	7342	2.6
Yemen	7341	3.0
Angola	6944	2.5
Tunisia	6630	2.3
Malawi	6507	3.0
Upper Volta	6360	2.0
Zambia	6330	3.1
Senegal	5991	2.7
Bolivia	5916	2.6
Dominican Republic	5813	3.0
Niger	5634	3.3
Guinea	5285	2.9
Rwanda	5276	3.4
Hong Kong	5233	2.4
Haiti	5195	1.7
Denmark	5123	0.3
Somalia	5116	2.8
El Salvador	5087	2.9
Finland	4818	0.4
Burundi	4778	2.2
Chad	4643	2.0

Table 7.4. (*cont.*)

Country	Number (mid 1982) $\times 10^3$	Growth rate (av. 1970–81) (%)
Norway	4133	0.5
Israel	4064	2.6
Honduras	3955	3.4
Puerto Rico	3952	–
Laos	3901	1.9
Sierra Leone	3672	2.6
Benin	3621	2.7
Ireland	3483	1.3
Libya	3425	4.1
Lebanon	3314	0.6
Paraguay	3251	2.6
New Zealand	3190	1.5
Papua New Guinea	3126	2.1

(Population data from *Encyclopaedia Britannica Book of the Year 1984*; growth rates from *World Development Report 1983*.)

Table 7.5. *Doubling times for different population growth rates*

Growth rate (% per year)	Doubling time (years)
0.1	730
0.2	350
0.4	180
0.6	120
0.8	90
1.0	73
1.5	49
2.0	35
2.5	28
3.0	24
3.5	21
4.0	18
4.5	16
5.0	14
5.5	13
6.0	12
6.5	11

only ones with rates above 1.0 were New Zealand (1.5), Australia (1.4), Ireland (1.3), Canada (1.2) and Spain (1.1).

Sources of resistance to fertility regulation

The mechanism referred to earlier which seems to underlie the reproductive patterns of *K*-selected animals can, of course, be inferred to function in both directions, that is to say, to favour reduction in fecundity in some circumstances and to resist reduction in others. The former action we have seen as playing a part in the demographic transition following a steady fall in the death rate if environmental conditions are appropriate; now we need to consider circumstances or human reactions that tend to work in the opposite direction, opposing any move to reduce the fertility rate of a population. Recognition of the existence of these processes can be important for the successful prosecution of national programmes for family planning; ten examples are discussed in the following pages.

Fig. 7.9. The likely proportions of the world population in various regions when it stabilizes around the year 2110. (After Rami Chhabra in *People*, **8**, no. 3, p. 28 (1981).)

High death rate

Clearly, a high death rate is a major influence, the common cause being starvation and intercurrent infection. Early in 1984, the UN Food and Agriculture Organization (FAO) reported that, as a result of drought, cattle disease (rinderpest) and the depredations of grain-boring insects, exacerbated by severe problems in the recruitment of aid, and food storage and distribution, about 20 million people faced starvation in Africa, tens of thousands having already died. Tragically, that is a familiar story in Africa, for the sub-Saharan belt is notorious for its droughts, and death so often stalks the land. In the developing world, there are several countries in the same predicament. Moreover, it is the children who are hardest hit: an Oxfam estimate in 1983 was that as many children as are *born* each year in the USA, UK, France, Sweden and Norway put together *die* in the Third World in their first year of life. Very high birth rates and total numbers are not surprisingly a 'natural' response to these circumstances (Table 7.6). In any event, if, as in many primitive societies, births are principally spaced out by lactation, infant death will automatically hasten the next birth.

Adverse social environment

Seemingly illogically, the highest birth rates are commonly found among the poorest sections of society, despite the fact that means of subsistence and general living conditions are so unfavourable. Thomas Doubleday said as much, back in 1842: 'There is in all societies a constant increase going on amongst that portion of it which is the worst supplied with food: in

Table 7.6. *Birth rate and infant mortality rate in relation to wealth*

Countries	GNP (US $ per capita)	Population × 10⁶	Birth rate per 1000 population	Infant mortality per 1000 births
Industrialized	5950	1350	16.2	15
Developing				
High income	4127	20	31.0	25
Upper middle income	1498	108	23.8	35
Intermediate middle income	721	370	41.4	48
Lower middle income	384	215	45.0	88
Low income	151	554	46.6	129
*Centrally planned economies**	2112	1480	17.8	25

* Excluding China.

(From *World Atlas of the Child*. World Bank; Washington, D.C. (1979).)

short, among the poorest.' Investigations have shown that this does not simply reflect ignorance and irresponsibility, but is a biological response. For the poor, children represent a kind of wealth: they can learn at an early age to help maintain the family by work, begging or theft, and they provide support to parents in sickness and old age. In a highly competitive system, a large family, sticking together, does better than a small one.

There is also the influence of long-standing attitudes in the community, particularly in underprivileged societies. A recent WHO study in the rural Western Province of Kenya revealed that large families are traditionally regarded with high approval, and most young people are brought up to understand that their proper function in society involves having many children. This is a highly fertile population with a growth rate of over 4 per cent, an average family size of 8 and a desired size of 12, but it is futile to try to persuade the women to use methods of family planning – they are not supposed to make decisions like that – their husbands would be dead against the idea and they would both feel keenly the opprobrium heaped on them by the community for such a course of action. Besides, these people live outside the money economy under very primitive conditions and are essentially at the mercy of the environment; planning of any kind is foreign to them, and family planning is certainly no exception.

Consistently, not only is a high birth rate positively correlated with the degree of national poverty (Fig. 7.10), but the expressed wish of couples for four or more children is also positively correlated with poverty (Fig. 7.11). So it is not surprising that the countries with the worst prospects of a rising GNP are among the most densely populated (Table 7.7).

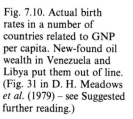

Fig. 7.10. Actual birth rates in a number of countries related to GNP per capita. New-found oil wealth in Venezuela and Libya put them out of line. (Fig. 31 in D. H. Meadows *et al.* (1979) – see Suggested further reading.)

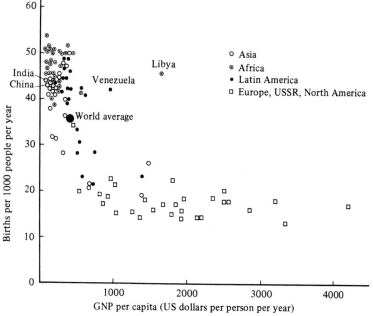

International rivalry

Obviously, the larger the nation (other things being equal), the greater the influence it has in all matters concerned with defence against aggressors, boundary disputes, control of other countries' affiliations, and access to important resources. There can clearly be simplistic but compelling reasons for a country to strive towards the largest population it can support without undue hardship for its citizens. Opposition to fertility control

Table 7.7. *Estimated GNP changes in selected countries*

	GNP (US $ per capita)	
Country	In 1968	Estimated for year 2000
China	90	100
India	100	140
USSR	1100	6330
USA	3980	11000
Pakistan	100	250
Indonesia	100	130
Japan	1190	23200
Brazil	250	440
Nigeria	70	60
West Germany	1970	5850

(From *The Limits to Growth*, see Suggested further reading.)

Fig. 7.11. Percentage of population expressing a desired family size of four or more, related to GNP per capita. (Fig. 32 in D. H. Meadows *et al.* (1979) – see Suggested further reading.)

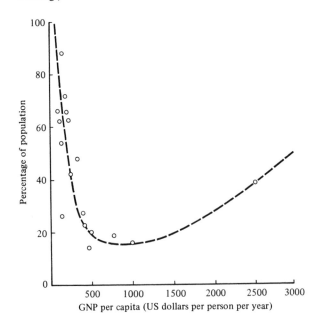

(pronatalism) can thus readily become a logical part of government policy, as it was for Hitler and Mussolini years ago and is for those in charge of Sudan and Malaysia today. Increasing expenditure on arms is a chilling feature of international affairs these days (Fig. 7.12), and this means less money available for maintenance or improvement of social conditions. Tragically, as Fig. 7.12 shows, the *rate of increase* in arms expenditure is greater in the developing world than in industrialized countries, and some countries now spend a good deal more on arms than on health (Table 7.8).

Increased life expectancy

With declining death rates and improving standards of health in the community, people live longer and an increasingly large proportion of the population survive into the post-reproductive years (see Fig. 3.1). As a result, the population tends to become 'top-heavy', the overburden representing the non-reproductive, non-productive, retired, aged and infirm, who are a drain on the national economy. Less well-off countries fare better if they can keep a preponderance of people of the 15–50 age bracket, who do most of the work and show most of the originality and drive, and understandably appreciate these short-term benefits of a pronatalist policy.

This line of thinking has in fact been acted upon on several occasions recently. Helmut Kohl offered tax incentives to coax West German housewives to produce an additional 200 000 children, the population having diminished steadily in the last 12 years. The governments of

Fig. 7.12. Military expenditure in the industrialized and developing worlds; the *rate* of increase is greater in the developing world. (From *New Internationalist*.)

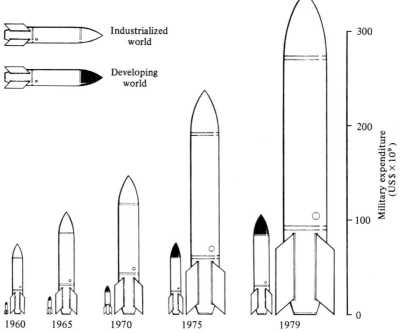

Hungary, USSR and Romania have followed a similar course. Fred Pearce writing in the *New Scientist* in 1984 (see Suggested further reading) supports this cause to great effect, pointing out that, among other things, even migrations of rural populations to cities, often regarded as catastrophic, can bring real benefit. He cites the International Planned Parenthood Federation (IPPF) as concluding from a recent study of Calcutta that 'Over the years migration has brought the city prosperity', and the World Bank has stated that 'Most studies conclude that migrants are assets to the urban economy.' Presumably it is all a matter of timing: if things happen too fast it is disaster, but at a slower tempo, or after a period of 'settling in', it could bring salvation, and so fuel pronatalist flames.

Table 7.8. *Comparison between the money spent on arms and that spent on health (in US dollars per capita)*

	Population (1980) × 10^6	GNP (1980)	Arms (1979)	Health (1979)
Low-income countries	2160.9	260	9	1
Nepal	14.6	140	1	1
Somalia	3.9	—	16	2
Burma	34.8	170	5	1
Upper Volta	6.1	210	4	1
Malawi	6.1	230	5	2
Sri Lanka	14.7	270	2	5
Tanzania	18.7	280	15	3
Middle-income countries	1138.8	1400	39	15
Kenya	15.9	420	13	5
Honduras	3.7	560	9	7
Thailand	47.0	670	15	3
Papua New Guinea	3.0	780	7	13
Morocco	20.2	900	35	7
Peru	17.4	930	17	8
Malaysia	13.9	1620	38	15
Brazil	118.7	2050	11	21
Industrialized countries	714.4	10320	283	235
New Zealand	3.3	7090	72	241
UK	55.9	7920	249	219
Finland	4.9	9720	92	197
Australia	14.5	9820	164	187
Canada	23.9	10130	131	126
USA	27.7	11360	376	183
France	53.5	11730	190	406

(Modified from *New Internationalist*, September 1983.)

Objectionable features of contraceptives

Essentially, all methods of contraception involving chemical agents or intra-uterine devices have their attendant health risks and objectionable side-effects, and these increase the longer they are used (see Chapters 2, 3 and 4). Because ill-effects may not be seen until after 20 or more years of use, it is only very recently that misgivings have come to be expressed seriously – after all, the Pill did not become generally available until the middle 1960s. Progressive improvements are being made, and will continue to be, but anxieties are hard to dispel and they readily spread, especially with the sensation-hungry public media. Commendable caution in the developed world creates needless alarm in the developing countries, where the benefits of contraceptives like the Pill far outweigh the disadvantages. Nevertheless, the WHO announced in their 1984 annual report (see Suggested further reading) that they have recently established a new Task Force on Safety and Efficacy of Fertility Regulating Methods, so the debate is not yet closed.

Some members of the public have reacted to the known drawbacks of existing contraceptive agents and devices by adopting 'natural' methods for fertility regulation, the most common depending on abstinence during the fertile period of the menstrual cycle. The difficulty is to detect the fertile period reliably, especially near the beginning and end of a women's reproductive life, when cycle length varies greatly. Even at other times, the failure rate of the method is rather high, and the net effect of all this is increased fecundity.

Problems of drug development

Producing and marketing a new contraceptive is a massive financial undertaking, more so than with a new drug intended for therapy, because the contraceptive is to be used on healthy subjects, usually continuously and for many years. To begin with there is the difficulty of finding a worthwhile 'lead'. Variations in the structure of known chemical compounds have effects on biological action that can very rarely be predicted, and it is virtually impossible to devise a new compound from theory for a particular action. Basic research therefore necessarily involves more or less random synthesis and testing of vast numbers of compounds, with minute chances of success with any. Many attempts have been made to pick up promising leads from traditional, herbalist medicine, but extraction of the plant material has almost always yielded only a medley of chemical entities with minor or irrelevant actions.

Once a new and promising chemical entity is recognized, it has to be tested on animals – usually rats to begin with – to see if it has the right kind of action and is not too toxic. Then other animals, including rabbits, beagle dogs and rhesus monkeys, are tried. If all goes well, the first trials on human subjects are then carried out, initially testing for toxicity, metabolic products, etc., and then for anti-fertility action. If this looks promising, it is back to the laboratory to devise economically viable methods of synthesis and manufacture to permit large clinical trials to be undertaken. These involve thousands of subjects, usually in several different countries, over a number of years. Usually, by this time something has proved unsatisfactory somewhere along the line and the whole process has to be repeated, so that when a drug can be offered on the open market, about 17 years have passed since the start, and the cost of development is about $30 000 000 (according to Carl Djerassi, writing in 1978, see Suggested further reading). Since 1978, standards of safety have been raised and the number of tests at all stages increased, so the time required and the cost involved must both be a good deal greater. It is not surprising, therefore, that chemical firms are increasingly reluctant to risk time and money in this way, and indeed much drug development work is now carried out with the financial help of non-profit-making organizations like the WHO, the Population Council and the National Institutes of Health.

Ideology

There are many religious and cultural bans on measures designed to limit normal fertility. For committed Roman Catholics, as well as for followers of many other creeds, the biblical injunction 'Be fruitful and multiply' (or its equivalent) is not to be dismissed lightly. Few have interpreted the phrase so faithfully as the Hutterites and the Amish, Anabaptist sects that originated with Jacob Hutter in Switzerland and Jacob Amman in Austria, and now survive in parts of the USA and Canada. Hutterite families in recent times *averaged* around 10 or 11 children, and the Amish between 8 and 9. When Eli Miller, a Hutterite farmer in Ohio, died at the age of 95, he left 410 living descendants. Different ideologies similarly propel Hasidic Jews and certain Islamic sects to achieve high prolificacy.

Need for a male child

Throughout human society there is a general preference for boys rather than girls, which often results in needlessly large families. The preference is attributable partly to the long-established practice of patrilineal inheritance, which requires that the family name and fortune go to the eldest son or the other sons, and partly to the fact that males are generally bigger and stronger than females and so capable of a heavier physical work-load. In Western countries, the preference is less clearly shown, and opinion polls have often revealed a fairly equal division of wishes.

If the sex of the future child could be controlled, that would have a major influence on fertility rates in certain developing countries, but the prospects at present are dim. In some more primitive and isolated communities, such as among the eskimo, among the South Sea Islanders, in remote parts of China and among some African tribes, the preference for a male leads to the practice of female infanticide, fortunately now increasingly rare.

Idiosyncrasy

Closely related to those who favour prolificacy in obedience to doctrinal directive are those who have a deep-rooted desire for a large family and hold it as their right to proceed accordingly without regard to any declared national population policy or implicit local custom. There are also certain feminist groups who object to the spearhead of contraceptive policy being directed at the female body. Those defending the 'right to life', and similar movements, which appear to be growing in America and Europe, draw strength from the prohibition against murder that exists in many religions, and accord the conceptus full rights, from the *beginning*. One comes up inevitably against the problem of competing rights, i.e. those of the few-celled human proembryo against those of the human being born into an already crowded world; nevertheless the net effect is growing opposition to contraception.

Another 'quirk' with a pronatalist influence is 'machismo' or its equivalent. The dictionary defines it as 'The need to prove one's virility by acts of daring or courage', but in Latin America it has come to signify pride in demonstrating one's success with women by generating numerous progeny. Needless to say, this activity, all too common in some areas, works actively against fertility control.

Equally idiosyncratic is the claim by adolescent sex enthusiasts that 'it is more fun without protection'. The age of puberty has unfortunately diminished steadily in developed countries during the past 100 or so years, and children may acquire their fertility before they develop a sense of social responsibility. We must accept that this trend will also overtake the developing world in due course, exacerbating its population problems.

Ignorance

Lack of a clear appreciation of the causative relation between intercourse and pregnancy is uncommon these days, though it used to be widespread among the native peoples of the Pacific area. Nowadays, the ignorance relates more to what measures to take to restrain fertility. Education in family planning techniques is urgently needed, but is difficult and expensive to introduce into remote parts of developing countries. This aspect of Social Services is proving widely acceptable in the Third World because its primary aim is birth spacing (and only incidentally birth reduction), which has important advantages for maternal and child health and does not conflict with a developing country's self-esteem or such pronatalist feelings as may exist in the community. Accordingly, it is an area that is receiving good support from organizations like the UNFPA and WHO, but there is a very long way to go before we can claim to have reached any realistic goals. Education is also needed in developed countries to provide contraceptive advice to young teenagers, but there is a great reluctance to do this and the consequence seems inevitable: in the USA, for example, about 300 000 teenage abortions are performed each year.

Positive feedback loops

If we look again at the assorted barriers to population control that have just been debated, it soon emerges that several of them tend to participate in positive feedback loops, which, by definition, represent trends that are not only generated by population increase but in turn generate population increase. Fig. 7.13 sets out this notion diagrammatically, showing four such loops.

In *Loop 1*, the growth of population increases the demand for food so that agriculture comes under increasing pressure. People plant more grain, introduce improved varieties of crop, use more fertilizers and pesticides, try for an additional harvest, induct more land by destroying forests, increase the animal stocking rate and introduce or increase mechanization. Some of these endeavours are successful, a few dramatically so, but success is too often only temporary and the list of failures woefully long. The reasons are varied. Heavily exploited land is extremely susceptible to the unexpected dry season; crops fail, and with nothing to bind the soil, wind and water erosion play havoc with the naked country. Often, previously forested land turns out to be unsuitable for agriculture owing to low fertility and mineral deficiencies, and again erosion depletes the invaluable top-soil, which will take many years to replace. Sometimes, in hilly areas, deforestation leads not only to erosion but to landslides, causing enormous damage. In addition to all this, increasing population in the countryside impels progressive subdivision of the land, the holdings eventually getting too small for subsistence.

As a consequence, the farming community is forced to the conclusion that making a living on the land is altogether too difficult, so they pack up and move to the towns and cities in the hope of finding work, and to benefit from overseas relief programmes which are more accessible in the

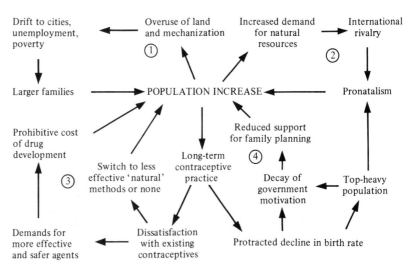

Fig. 7.13. Positive feedback loops that tend to maintain population increase.

distribution centres. The drift to the cities is a world phenomenon and will clearly lead to the development of truly enormous conurbations by the end of this century (Fig. 7.14); ironically, these will be greater in the developing world than elsewhere, their growth will be far too rapid for proper organization and management, and huge slum areas will develop, as they already have in Calcutta, Bombay and Mexico City. These provide the adverse social environment that we noted above as the second major source of resistance to fertility regulation, leading to larger families and favouring population increase.

Loop 2 concerns the role of population growth in the drive to ensure a supply of natural resources that are internationally most in demand. This is not to say that these resources are necessarily finite and will eventually become exhausted – I will discuss this point more fully later – but resources may be in more urgent need or wider demand than can be supplied, or the route of supply for one country may become subject to restrictive control by another, or the resource reserves may be *thought* to be approaching exhaustion. In any of these circumstances, with their obvious bearing on competing interests between nations, those that have a falling birth rate and an ageing top-heavy population will be at a serious disadvantage. Pronatalism then readily becomes government policy and with it firm opposition to fertility control. One can question the actual efficacy of governmental pronatalist programmes when social conditions

Fig. 7.14. Populations of major world cities. (Data for 1950 and 1975 from *New Internationalist* and for 2000 (estimated) from *Estimates and Projections of Urban, Rural and City Population: the 1982 Assessment,* UN Population Division (1985).)

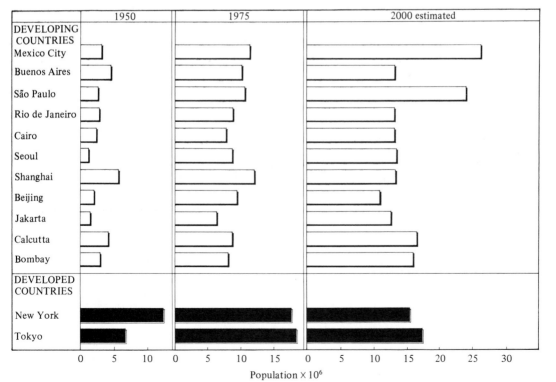

are at a good level and people are becoming more materialistic and likely to regard children as an economic liability, but many communities in the Third World have not entered this phase yet and are consequently fair game for pronatalist propaganda.

Loops 3 and 4 have a common origin among people who have had long-term use of contraceptives. One result (Loop 3) is increasing dissatisfaction with available methods for their health risks and other disadvantages. Some individuals will decide to switch to 'natural' methods, which in their present state of development are much less reliable; consequently, more people are likely to become pregnant and so help the population increase. Others will stick to their guns but clamour for new and better agents, the search for which is moving into a prohibitively expensive zone owing to the increasingly stringent requirements of drug regulatory authorities.

In the alternative reaction (Loop 4), the decline in birth rate resulting from long-term contraceptive use leads to change in government policy, because the goal seems to have been achieved or because the birth rate has dropped too low, and consequently there is reduced support for family planning programmes. In this connection, note the declining contributions to the WHO Special Programme of Research, Development and Research Training in Human Reproduction, which is representative of the international drive to stem the population flood (Fig. 7.15).

Consequences of population growth

Although a dramatic slowing in the overall world population increase is possible, such that equilibrium could be reached at about 8000 million near the start of the twenty-second century, it will need a long-term world-wide

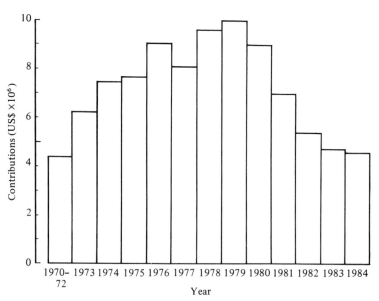

Fig. 7.15. Total contributions from various countries for the support of the WHO Special Programme of Research, Development and Research Training in Human Reproduction. (Figures obtained from Annual Reports and adjusted on the basis of inflation at the rate of 5 per cent annually.)

dedication to the policy of fertility control and the establishment generally of appropriate economic and social conditions. What is much more likely is that a more moderate course will be pursued and that the rate of increase will conform approximately with the figures set out in the second or third columns of Table 7.1, with plateaus of about 10000–12000 million, reached around the middle of the twenty-second century. Or it could even mean a plateau of about 14000 million near the end of the twenty-second century, unless, of course, an environmental limiting factor comes into operation before those times.

Does the prospect of a world population of 14000 million have to be an intolerable or even a dismal one? Let us examine some of the implications. Sheer living space would probably not be a serious problem. Hong Kong in 1981 boasted 4986560 people in a total land area of 1037.2 km^2; most of the people lived in the city which accounted for only about 4 per cent of the total area, the remaining space being more or less open country, so the concentration in the city was roughly 120000 people per square kilometre and, in the colony as a whole, roughly 4800 people per square kilometre. At the same *overall* population density the UK could contain 1200 million persons and the USA 45000 million, mostly in cities and towns. Major problems of food distribution, transport, sewage disposal, health services, social security and the maintenance of law and order would have to be met, but technology is likely to advance rapidly and time will become less pressing as the population growth curve flattens out.

Food production

With millions of people now suffering from malnutrition, ranging from relative protein and vitamin deficiencies to outright starvation, food production might well seem a good candidate as the ultimate limiting factor, but it need not be so. In 1975, Kenneth Mellanby (see Suggested further reading) calculated that the population of the United Kingdom (then about 54 million) could be provided with a nutritionally adequate diet (consisting mainly of potatoes, sugar, milk, meat and cereal) on about half of the 19 million hectares of agricultural land available at that time. Since then, important improvements have been made in plant and animal production methods, so the arithmetic must be even more favourable now. The UK is high in the league of crowded countries and, although growing slowly – it takes 700 years to double the population (see Table 7.5) – it is losing agricultural land at the rate of some 20000 hectares a year through expansion of what we can call 'people space', namely urban areas, road systems, forests maintained as public amenities or economic resources, airports and airfields, national parks and conservation areas, dams and freshwater reservoirs, open-cast mines, rubbish dumps, sewage treatment works, nuclear power stations, and so on. However, even at this rate of loss, it will take about 400 years to use up the present 'surplus'. And in that time, at present undreamed-of advances in methods of food production will take place. So, on this rough calculation, we can surmise that food production in the UK, and in many countries similarly placed in the developed world, should not only suffice but even allow large exports for many years to come, long after the world's population has reached equilibrium.

Of course, conditions in the developing world are very different, and in many areas food production is frequently arrested by drought. In September 1983, FAO reported 33 countries suffering food shortages. Particularly afflicted were areas of sub-Saharan Africa, conditions being worsened by political strife and economic difficulties, and the official estimates of necessary food imports approached 8 million tonnes. With rapidly expanding populations there, the future prospects would appear gloomy indeed. The FAO, however, are more optimistic, maintaining that many African countries have ample land resources awaiting development. Similarly, in Latin America and Asia, there is great scope for increased production with the use of modern methods. In 1984, FAO estimated that, with efficient farming, world agriculture could provide enough food for 33 000 million people. Countries vary a great deal, though, and some are likely to remain dependent on outside aid for a long time yet. Collaborative effort between members of the developing community would assist enormously in the solution of world food problems, countries with more than adequate agricultural land coming to the aid of the needy. As FAO point out, even with present-day production methods, the Congo could feed 20 times its present population and Gabon 100 times its population.

These assertions stand in stark contrast to the predictions made by the Club of Rome (see Suggested further reading), who maintained that the present rapid increase in the extent of total world agricultural land would come to an abrupt halt early in the next century owing to the expansion of 'people space' (Fig. 7.16). This component (the nature of which was discussed earlier) must certainly grow proportionately with the population, and so, from the time the two curves meet, the area of land available for food production must *decrease* exponentially. Has this effect been taken into account by FAO? If our food-producing capability were indeed closely dependent on land area, then such a turn of events must surely be ominous for the human race in the not-too-distant future.

But does *area* really limit food production on land? As we have seen, it has not in the United Kingdom, where improvements in production methods have more than compensated for both increasing consumer demands and loss of land. The rest of the EEC seems likely to be similarly placed, to judge by the present potential for creating butter and sugar 'mountains', milk 'lakes', etc. Clearly, though, the traditional approach to food production cannot be expected to succeed indefinitely, especially after 'people space' starts to bite large chunks out of agricultural land on a world scale. The technology must change, rapidly and dramatically.

It can do so, too, given the proper funding, application and effort. There are all sorts of ways; let us take a few examples. Already, in Europe, the increasingly widespread use of glass and plastic greenhouses has enormously improved the volume and variety of vegetable production; even greater yields are obtained by incorporating hydroponics. The 'greenhouse' approach to fresh-water fish production in a cold climate has also been singularly successful. As Peter Kalinowski showed 10 years ago (see Suggested further reading), it is perfectly feasible to grow carp up to about 0.5 kg in weight each in one season, when they are kept in a greenhouse at a concentration of up to 230 kg of fish per cubic metre in a tank with closed circulation but an efficient pumping and gravel-bed sewage-treatment facility, and fed an appropriate diet. No one in his right mind would wish for a steady diet of carp, but the fish do represent high-class protein and essentially the same methods can be applied to more palatable species at rather lower crowding rates. In warm climates, outdoor fish farms and sea-water enclosures are already producing vast amounts of fish protein, and much more besides could yet be done.

Single-cell protein has interesting possibilities and its production is even less demanding of space. This involves varieties of algae, yeasts, bacteria, some fungi and perhaps protozoa. A green alga, *Scenedesmus obliquus*, has been well investigated; it is readily grown on open water and harvested by skimming off and drum drying in a process like that used to produce powdered milk. The powdered algae can be added to a basic ration of cereal or rice boosting the calorie, protein and vitamin A, B and C content considerably. Work on yeasts and bacteria is more at the research stage,

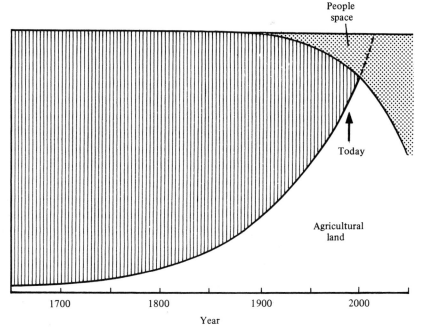

Fig. 7.16. Total area of the world arable land (vertical shading) and its progressive utilization as agricultural land (white) or for 'people space' (stippled). (Based on Fig. 10 in D. H. Meadows *et al.* (1979) – see Suggested further reading.)

except, of course, for brewers' yeast *Saccharomyces cerevisiae*, which was used as a 'meat extender' during the Second World War. More recently, yeasts of the genus *Candida* have drawn particular attention. To produce yeasts and bacteria, nutrient substrates are needed, but these can be any from a wide range of materials, including petroleum wastes, molasses, starch, sulphite waste from paper manufacture, cheese whey and cellulose. Problems have still to be solved, including those relating to pathogenicity, toxic metabolites and allergenicity, but the production of single-cell protein in tanks would make a relatively insignificant claim on land area.

Eventually, it should prove possible to synthesize nutritional requirements (except minerals) from elemental components, either with the aid of 'engineered' yeasts and bacteria – lysine, threonine and glutamic acid can already be produced in this way – or directly, with the aid of boundless energy from nuclear fusion.

Natural resources

Our natural resources are commonly regarded as either renewable or non-renewable, the latter giving rise to the greater concern. Fossil fuels seem undeniably non-renewable, but what is not known is their full extent or whether they will always be of critical importance. Minerals are non-renewable in their natural state, but most are important for the metals they contain and these are not destructible; they simply change into less immediately usable forms or pass into less accessible regions, such as into the oceans from which recovery requires currently expensive technology. The relative scarcity of minerals is judged by their ease of procurement, as well as by the magnitude of the demand for them, and estimates of 'world reserves' are really just measures of existing technical proficiency. Fig. 7.17 shows how much estimates of 'known reserves' can change with time.

Possibly the most threatened natural resource of an inorganic nature is fresh water; there's no shortage of the salted variety! Especially to be prized is *clean* fresh water, a commodity treated with scant respect in much of the developed world but often a matter of life and death in the developing world. Potable (i.e. palatable and safe) water is something that people in certain lands consider worth travelling miles to get, spending a great deal of physical effort and a significant part of each day doing so. And potable water seems bound to get scarcer as the population grows. Indeed, the provision of fresh water is one of the major problems for the immediate future in developing countries, especially those in the tropical and sub-tropical belts, where most developing countries lie.

One of the main practical things that can be done to conserve fresh water is to dam a river and make a reservoir. This provides not only a source for general use, but also a head of water for irrigation and hydroelectric power, from which water can be released at a controlled rate, thus converting the old flood-and-drought pattern into a steady moderate supply on which farmers and townsfolk can depend. There were 54 major dams under construction in 1982 and 53 in 1983, most in developing countries. But there are serious drawbacks to river damming. For one thing, the artificial lake created covers a large area of country which, being near a river, was almost certainly inhabited and cultivated, or else rich in wildlife, so its development causes great disturbance to the local economy and conservation, and records show that the human communities affected are rarely properly compensated for their losses. Then there are circumstances in which seasonal flooding of the neighbouring countryside served a useful purpose, as in the case of the Nile: the organically rich silt freshly deposited on the fields along the riverside was what the people used to depend on for growing their crops. And dams are not forever – many have been choked with water hyacinth or filled up with sediment, especially

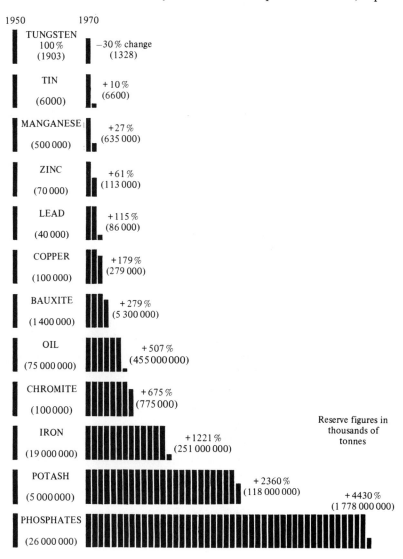

Fig. 7.17. 'Known' reserves of selected metals and minerals in 1950 and 1970. (Fig. 2.1 in J. L. Simon (1981) – see Suggested further reading.)

1950   1970

TUNGSTEN
100%   −30% change
(1903)   (1328)

TIN   +10%
(6000)   (6600)

MANGANESE   +27%
(500 000)   (635 000)

ZINC   +61%
(70 000)   (113 000)

LEAD   +115%
(40 000)   (86 000)

COPPER   +179%
(100 000)   (279 000)

BAUXITE   +279%
(1 400 000)   (5 300 000)

OIL   +507%
(75 000 000)   (455 000 000)

CHROMITE   +675%
(100 000)   (775 000)

IRON   +1221%
(19 000 000)   (251 000 000)

POTASH   +2360%
(5 000 000)   (118 000 000)

PHOSPHATES   +4430%
(26 000 000)   (1 778 000 000)

Reserve figures in thousands of tonnes

following deforestation of the catchment areas – particularly in the tropics, where a dam's lifespan may be as short as 30 years. Finally, river damming can seriously affect the lakes or inland seas into which they flow. For example, the level of the Caspian Sea has dropped 2.4–2.7 m since 1929, owing to damming and irrigation projects in the Volga and Ural Rivers, with consequent disorganization of coastal life and industry (seriously interfering with caviar production!).

Another way to help alleviate the water problem is to dig wells, and this device has supported rural communities from time immemorial, but there are problems here too. If the well taps a spring, it might indeed supply good water almost indefinitely, but if it taps the sub-surface water basin of a district, there are at least three difficulties: the water could have a high salt content and even if potable be unsuitable for irrigation, large-scale withdrawal could threaten the plant life of the area, particularly the trees, and long-term withdrawal could cause land subsidence. Much depends on the quantity of water used, and of course this in turn depends on the size of the dependent population.

Among the natural resources, the forests deserve special mention, for timber is critically important, as a commodity that can be exported to earn a developing country hard currency, as the most adaptable building material and (perhaps most critically) as a fuel. Many primitive communities have no alternative but to burn wood for personal warmth, for cooking and for boiling water to make it safe to drink. But the forests of the world are being destroyed at the rate of about 70 hectares a minute, day in, day out, and vast regions have now become quite depleted. In some parts of Africa there is only enough firewood left to allow people to have one cooked meal a week. The rate of deforestation enormously exceeds the rate of replacement. A particularly vivid account of conditions in Nepal appeared in the *Observer* newspaper on 29 May 1983, and extracts are worth quoting verbatim:

> The Himalayas are disappearing into the sea. Tons of soil are being stripped from every acre of the mountains by each monsoon, destroying the land on which Nepal's people depend. The soil is swept down the Ganges river system and is forming a new island in the Bay of Bengal. This disaster has developed with extraordinary swiftness. In the three decades since Everest was conquered half of the forests that once coated Nepal's mountains have been cut down and the rest are expected to disappear by the end of the century. The trees bound the soil to the mountain slopes and enabled it to soak up the fierce monsoon rains, releasing them gradually to enrich the whole region. Now the rains are a curse, not a blessing. Each year they wash away more than 12 tons of soil from every acre of the bare hillside. In the worst areas, 80 tons of soil are ripped from each acre. Villagers watch all night for landslides during the monsoons. Little wonder – 20000 landslides have been recorded in a single day, sweeping away the

terraces of fields carved laboriously into the mountainside and burying whole villages and their inhabitants. As the soil goes, crops fail. Rice yields have dropped by a fifth in just 5 years in the hills, maize by a third. In many areas the once-tall maize now grows no higher than millet. Already the people are hungry. The average hill farmer can grow only enough food to feed his family for 8 months of the year, and 60 per cent of the children are stunted by famine.

But the plains are dangerous, too. The mountain soil coats the river beds, raising them 6 to 12 inches a year. Every year the rains pour more forcefully off the bare hillsides into shallower channels. Inevitably the rivers flood and the land on their banks is swept away. The erosion of the Himalayas has greatly increased flooding in the Ganges basin, one of the world's most densely populated and important food-growing areas. In all, 300 million people are affected by what goes on in the mountains.

Pollution

Opening the topic of pollution and Pandora's box are rather similar acts in terms of consequences. There seems to be no limit to the variety of ways in which man can despoil his planet, and varying levels of local or international concern have been expressed. We now have the production of 'acid rain' which is destroying crops and forests in North America and parts of Europe; the discharge into the Rhine of salt from French potash mines in Alsace; the industrial pollution of the Elbe in East Germany and Czechoslovakia; serious pollution in the Bosporus and the Golden Horn by untreated sewage, oil, market refuse, industrial wastes and other debris; the pumping of 2000 million litres a day of toxic wastes, including dioxin, into the Niagara River in New York State; the contamination of some 10000 private wells in California with dibromochloropropane, a pesticide banned from public use 5 years previously; the contamination of 97 per cent of the population of Michigan with polybrominated and polychlorinated biphenyls and DDT; the contamination of soil and garden produce by lead and cadmium residues from old Roman mine workings in the village of Shipham, Somerset, UK; as yet unsuccessful moves to ban the addition of lead to petrol in most Western countries; the prediction that 8000 workers in Canada and the USA would die annually as a result of exposure to asbestos until the manufacturing use of the mineral ceased; the recurrence of the serious smog levels in Athens, resulting in major restrictions on factory production and motor traffic; the problem of smog also in Ankara, which required schools to be closed and car traffic restricted; the possibility that ozone, produced chiefly by motor vehicles, might be causing crop losses amounting to $3.1 million a year in the USA; the dumping of radioactive waste in the Atlantic by the UK Atomic Energy Authority.

Most of the events described are the products of highly industrialized societies, but in due course they will come to be shared by at present developing countries. Already, some, like Taiwan, are beginning to have serious pollution problems; others have accepted toxic wastes from developed countries in order to earn hard currency, although they lack proper means for disposal. Probably the chief pollutant contributed by the developing countries at present is CO_2 from wood burning; ironically, it is they too who are mainly responsible for the destruction of forests, which constitute the most important balancing factor through the absorption of CO_2. But the developed countries must take much of the blame, for their heavy demand for various kinds of timber sustains that market, and they have by far the larger share of fossil-fuel-fired industrial machinery and internal combustion engines. Inevitably, atmospheric CO_2 levels are increasing exponentially (Fig. 7.18), which is considered likely to augment the 'greenhouse effect' of the earth's atmosphere so that world climates would gradually get warmer. In 1983, the US Environmental Protection Agency estimated a doubling of pre-industrial CO_2 levels and an average temperature increase of 2 °C by the year 2040, and a temperature increase of 5 °C by 2100. In response, the US National Academy of Sciences expressed the opinion that the rise would be between 1.5° and 4.5 °C by the year 2100. Man himself and his agriculture generally may well be able

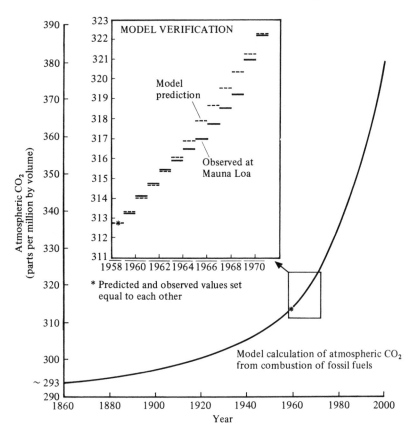

Fig. 7.18. Progressive increase in atmospheric CO_2 since 1860. (From Fig. 15 in D. H. Meadows *et al.* (1979) – see Suggested further reading.)

to adapt to such temperature changes, but these are not without further implications. The higher temperatures would eventually reduce the mass of polar ice caps and so raise ocean levels, flooding low-lying coastal areas, including towns and cities. A truly 'science fiction' prospect!

There is no doubt that world population will continue to increase rapidly, possibly exponentially, for the immediate future. Later the pace will slacken and population numbers will tend to stabilize, but how much later and at what level depends a good deal on how world affairs are managed.

Opinions in recent years have become more divided as to the benefit or burden a high birth rate can be judged to confer on a country, but there *is* consensus that purposive control of birth rate – family planning – is a worthy aspiration. For most developing countries this means restraint to allow time for economic policies to mature and bear fruit, and the aim of these policies is most logically directed at securing a social and political environment in which people are free to determine a family size in keeping with the living standards they find acceptable. But these laudable ambitions are often confused and even thwarted by domestic strife or international rivalry, by concern over health risks of contraceptive measures, by conflicting ideologies, by idiosyncrasies and by ignorance – in short, by the barriers that exist in human nature against population control.

Suggested further reading

Time bomb or myth: the population problem. R. S. McNamara. *Foreign Affairs*, **62**, 1107–31 (1984).

In defence of population growth. Fred Pearce. *New Scientist*, pp. 12–16. 9 August (1984).

Recirculating systems for coarse fish rearing. P. M. Kalinowski. In *Seventh British Coarse Fish Conference*, University of Liverpool, March 1975. Janssen Services; Billingsgate, London (1975).

North–South: a Programme for Survival. The report of the Independent Commission on International Development Issues, under the Chairmanship of Willy Brandt. Pan Books; London and Sydney (1981).

People, vol. 11, no. 4. Report of the International Conference on Population, Mexico City, August 1983. Longman; London (1984).

World Development Report. The World Bank. Oxford University Press (1983 and 1984).

World Health Organization, Special Programme of Research, Development and Research Training in Human Reproduction. Thirteenth Annual Report. WHO; Geneva (1984).

Cataclysm: the North–South Conflict of 1987. William Clark. Sidgwick & Jackson; London. (1984).

The Politics of Contraception. Carl Djerassi. Norton & Co.; New York and London (1979).

The Social and Environmental Effects of Large Dams. E. Goldsmith and N. Hildyard. Wadebridge Ecological Centre; Camelford, Cornwall (1984).

The Limits to Growth; a Report for the Club of Rome's Project on the Predicament of Mankind. D. H. Meadows, D. L. Meadows, J. Randers and W. W. Behrens III. Pan Books; London and Sydney (1979).

Can Britain Feed Itself? Kenneth Mellanby. Merlin Press; London (1975).

The Ultimate Resource. Julian L. Simon. Robertson; Oxford (1981).

The Resourceful Earth. Ed. J. L. Simon and H. Kahn. Basil Blackwell; Oxford (1984).

Africa in Crisis. Lloyd Timberlake. Earthscan.

INDEX